Nanofluidics

Nanoscience and Nanotechnology

RSC Nanoscience & Nanotechnology

Series Editors

Professor Paul O'Brien, *University of Manchester, UK*
Professor Sir Harry Kroto FRS, *University of Sussex, UK*
Professor Harold Craighead, *Cornell University, USA*

This series will cover the wide ranging areas of Nanoscience and Nanotechnology. In particular, the series will provide a comprehensive source of information on research associated with nanostructured materials and miniaturised lab on a chip technologies.

Topics covered will include the characterisation, performance and properties of materials and technologies associated with miniaturised lab on a chip systems. The books will also focus on potential applications and future developments of the materials and devices discussed.

Ideal as an accessible reference and guide to investigations at the interface of chemistry with subjects such as materials science, engineering, biology, physics and electronics for professionals and researchers in academia and industry.

Titles in the Series:

Atom Resolved Surface Reactions: Nanocatalysis
PR Davies and MW Roberts, *School of Chemistry, Cardiff University, Cardiff, UK*

Biomimetic Nanoceramics in Clinical Use: From Materials to Applications
María Vallet-Regí and Daniel Arcos, *Department of Inorganic and Bioinorganic Chemistry, Complutense University of Madrid, Madrid, Spain*

Nanocharacterisation
Edited by AI Kirkland and JL Hutchison, *Department of Materials, Oxford University, Oxford, UK*

Nanofluidics: Nanoscience and Nanotechnology
Edited by Joshua B. Edel and Andrew J. deMello, *Department of Chemistry, Imperial College London, London, UK*

Nanotubes and Nanowires
CNR Rao FRS and A Govindaraj, *Jawaharlal Nehru Centre for Advanced Scientific Research, Bangalore, India*

Visit our website at www.rsc.org/nanoscience

For further information please contact:
Sales and Customer Care, Royal Society of Chemistry, Thomas Graham House, Science Park, Milton Road, Cambridge, CB4 0WF, UK
Telephone: +44 (0)1223 432360, Fax: +44 (0)1223 426017, Email: sales@rsc.org

Nanofluidics
Nanoscience and Nanotechnology

Edited by

Joshua B. Edel and Andrew J. deMello
Department of Chemistry, Imperial College London, London, UK

RSC Publishing

ISBN: 978-0-85404-147-3
ISSN: 1757-7136

A catalogue record for this book is available from the British Library

Published by The Royal Society of Chemistry,
Thomas Graham House, Science Park, Milton Road,
Cambridge CB4 0WF, UK

Registered Charity Number 207890

For further information see our web site at www.rsc.org

Preface

As any self respecting nanotechnologist knows, the 29th December 1959 was a rather significant date. Richard Feynman's address, to an audience of scientists and engineers, at the annual meeting of the American Physical Society, did not provide any quick fixes to the problems associated with "*manipulating and controlling things on a small scale*". Rather, Feynman's prescience and foresight, although built on established scientific principles and technology, provided a glimpse of a future world filled with a range of *nano-tools* that could have vast utility in chemistry, biology, engineering and medicine.

On reading Feynman's "*invitation to enter a new field of physics*" almost half a century later it is quite unnerving to see how close his predictions were to the mark. Although, certainly not perfect, his premise that the ability to manipulate matter on an atomic scale would facilitate new opportunities has certainly become reality. Most notably, Feynman was captivated by the possibilities of miniaturising computer circuitry and creating improved electron microscopes. As we now know, these ideas have since been brought to fruition. Microelectronic systems have shrunk to sizes approaching the molecular level and the development of scanning probe microscopes (e.g. STM and AFM) enable us to image and manipulate individual atoms.

At the heart Feynman's message is the idea of miniaturisation. The intervening years have seen many new and exciting developments using this simple concept. Of particular note has been the use of miniaturisation in solving chemical and biological problems. For example, microfluidic or lab-on-a-chip technology has taken much inspiration from integrated microelectronic circuitry. In simple terms, microfluidics describes the investigation of systems which manipulate, control and process small volumes of fluid. Development of microfluidic technology has been stimulated by an assortment of fundamental features that accompany system miniaturization. These features include the ability to process and handle small volumes of fluid, enhanced analytical performance when compared to macroscale systems, low unit cost, small device footprints, facile process integration and automation and high analytical throughput. Although, in many ways microfluidics takes advantage of the "*smaller, cheaper, faster*" paradigm from the microelectronics industry, the drive to smaller and smaller feature dimensions has not really been dominant in defining research avenues. In large part this is due to the obvious need for sufficient analyte to be present within the system but also is due to the fact that the interactions between a fluid and the walls of the microfluidic environment become increasingly dominant, and often problematic, as dimensions are decreased.

In a broad sense, the field of nanofluidics has developed, not as an extension and improvement of microfluidic systems, but rather as a way of exploiting certain unusual physical phenomena that simply do not exist at larger length scales. As is discussed in detail in this book, friction, surface tension, and thermal forces become increasingly dominant when feature dimensions become comparable to the size of the molecules or polymers contained within, and accordingly such *nanofluidic regimes* offer new opportunities for the manipulation of molecular systems.

In Chapter 1, *Cees Dekker* and colleagues provide a detailed discussion of the most fundamental nanofluidic structures; nanochannels. Both theoretical and experimental studies of the transport of molecular and biological species through such structures demonstrate the strong departure from bulk behaviour in nanoscale environments, and lay the foundation for how we might create new nanofluidic applications. In Chapter 2, *Jongyoon Han* provides a personal and fundamental discussion of significant engineering issues faced by nanofluidic technologists, with a particular focus on molecular separation, concentration and detection. Chapter 3 develops this discussion to provide an understanding of hydrodynamic flow fields in nanofabricated arrays of obstacles. Importantly, *Jason Puchella* and *Bob Austin* elegantly analyse some of the unexpected (but recurring) elements of flow in arrays of this kind. In Chapter 4 *Paul Bohn* and co-workers extend the discussion of flow in nanofluidic systems, but more specifically address the construction and operation of hybrid microfluidic-nanofluidic architectures. The authors clearly show how such hybrid systems may be used to solve some of the problematic issues currently faced in chemical and biological analysis, and also highlight the fact that integration of microfluidic and nanofluidic elements results in behaviour not observed in either system independently. Chapter 5 focuses on the use of nanofluidic elements as basic tools in modern day analysis. Specifically, *Jun Kameoka* and associates, examine a range of methods for fabricating nanofluidic conduits, and then demonstrate how such environments can be used to perform ultra-high efficiency single molecule detection. This theme is continued in Chapter 6 where *John Kasianowicz* and colleagues present the rationale for using nanometer-sized pores (rather than channels) to characterize biological macromolecules and polymer molecules. In particular the authors discuss the use of biological pores (such as α-hemolysin) in the design of efficient structures for DNA fragment sizing and separation, and speculate on future applications in biosensing, nanofiltration and immunoisolation. The use of nanopores in biological sensing is further expounded by *Joshua Edel* and colleagues in Chapter 7. In this contribution, the focus is on the potential utility of solid-state nanopores for single-molecule analysis and DNA sequence analysis. The authors highlight the flexibility of solid-state formats and propose a powerful new class of nanofluidic devices that allows for ultra-high throughput measurements at the single molecule level. In Chapter 8 *Li-Jing Cheng* and *L. Jay Guo* introduce the concept of Ionic rectification, a unique effect observed in nanofluidic devices. Importantly, the phenomenon of rectification relies on electrostatic interactions between ions and the fixed surface charges within a nanochannel, and thus may be used for the separation and detection of charged molecules. Finally, in Chapter 9 *Yoshinobu Baba* and co-workers review and discuss recent studies that utilise nanopillars and nanoballs for efficient DNA size separation.

We would like to express our sincerest thanks to all the authors for accepting our invitations to contribute to this book. The field of nanofluidics is still in its early stages of existence, however this is an exciting time, with a diversity of advances being made on many fronts. All the contributing authors are pioneers within the field and we are delighted to be able to showcase their collective endeavours in one volume for the very first time.

Our job as editors has been made significantly easier by the proof-reading skills of Shelly Gulati, Katherine Elvira, Fiona Pereira, Mariam Ayub, and Andrea Laine. Moreover, the striking cover image depicting the transport of DNA molecules through a solid-state nanoporous membrane was created by Murray Robertson, an artist with a rare ability to visualise ideas in science. In this case a picture is certainly worth a thousand words!

We hope you find this book a valuable source of information and insight into the growing field of nanofluidics.

Andrew James deMello & Joshua Benno Edel
South Kensington, London

Contents

CHAPTER 1

Transport of Ions, DNA Polymers, and Microtubules in the Nanofluidic Regime

DEREK STEIN[1,2], MARTIN VAN DEN HEUVEL[1], AND CEES DEKKER[1]

1 Kavli Institute of Nanoscience, Delft University of Technology, Delft, The Netherlands

2 Present address: Physics Department, Brown University, Providence, RI, USA

1.1 INTRODUCTION

Lab-on-a-chip fluidic technology takes inspiration from electronic integrated circuits, from which its name is derived. Lab-on-a-chip systems aim to improve chemical and biological analysis by using chip-based micromachining techniques to shrink the size of fluid handling systems.[1] In this way it borrows both the fabrication technology and the "smaller, cheaper, faster" paradigm from the integrated circuit industry. For silicon-based electronics, miniaturization eventually gave rise to qualitatively different transport phenomena because the device dimensions became comparable to important physical length scales, such as the de Broglie wavelength. Nanoelectronics has consequently become nearly synonymous with quantum mechanical effects. As fluidic devices are shrunk down to the nanoscale in the quest to manipulate and study samples as minute as a single molecule, it is natural to ask, "What physical phenomena should dominate in this new regime?"

As early as 1959, Richard Feynman recognized the challenges to controlling the motion of matter at the nanoscale in his famous speech, "There's plenty of room at the bottom".[2] He drew attention to the friction, surface tension, and thermal forces that would become important at such small dimensions. In the earliest nanofluidics experiments, the pioneering groups of Austin and Craighead observed unusual transport properties of DNA.[3-5] Channel dimensions comparable to the coil size of the polymers, called the radius of gyration, gave rise to strong entropic effects. Nanofluidics is in fact a regime where multiple physical length scales and phenomena become important, including the persistence length of a polymer, the Debye screening length for electrostatics, and the charge density along a channel surface.

In this chapter we review our studies of nanofluidic channels. These are the most fundamental structures in lab-on-a-chip devices, and represent the "wires" in the circuit analogy. It has therefore been natural to focus on the transport properties of nanofluidic channels, which we have investigated for small ions, DNA polymers that possess many internal degrees of freedom, and microtubules that undergo motion as part of their biological function. A recurring theme in our experiments has been the strong departure from bulk behaviour in sufficiently small channels. Different fluidic, statistical, or electrostatic effects can drive the crossover to a new regime in each case. This highlights the importance of understanding multiple interacting phenomena as new nanofluidic applications are sought.

1.2 IONIC TRANSPORT

Ions are ubiquitous in aqueous solution, and manifestations of their motion have been the subject of inquiry for centuries. In recent years the transport of ions in nanoscale systems has attracted increasing attention because of its importance to fundamental biological processes, e.g. ion channels in cellular and sub-cellular membranes,[6] as well as man-made porous membranes for applications such as fuel cells,[7] and solid-state nanopores for single molecule DNA analysis.[8,9] The motion of ions is also coupled to the motion of the fluid by viscosity. This gives rise to electrokinetic effects such as electro-osmotic flow (EOF), which is widely applied in lab-on-a-chip technology.[10,11]

In order to study the transport of ions in the nanofluidic regime in detail, we fabricated channels with highly controlled geometries that were straightforward to analyze using theoretical calculations. A typical slit-like channel is illustrated in Figure 1.1. The 4 mm long, 50 μm wide channel was lithographically patterned between two 1.5 mm x 2 mm reservoirs on a fused silica substrate. A reactive ion plasma then etched the fused silica at a rate of 30 nm/min and was timed to stop when the desired channel height, *h*, had been reached. The channels were sealed by bonding them to a second, flat, fused silica substrate. Bonding was achieved using either a sodium silicate adhesive layer,[12] or by direct thermal bonding.[13] Pre-drilled holes allowed access to the reservoirs for introducing fluids or electrical connections.

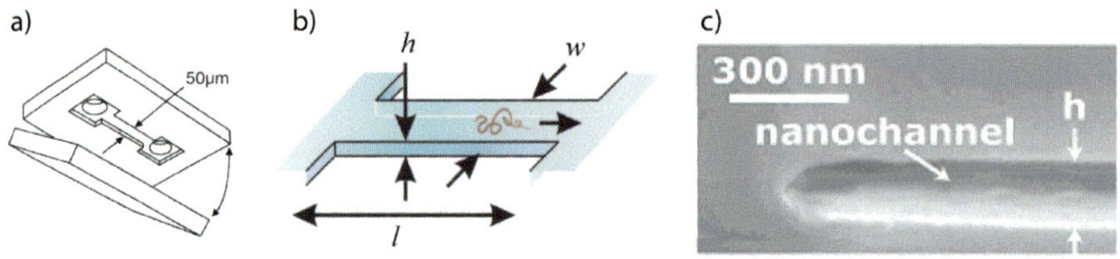

Figure 1.1 Slit-like nanochannels for transport measurements. (a) Nanofluidic channels are fabricated by bonding a flat, fused silica chip to a chip with a patterned channel structure and access holes. (image from ref.[14]) (b) The inner channel dimensions are well defined so that transport measurements of ions or polymers can be easily modeled theoretically. The channels are slit-like, with $l >> w >> h$. (image from ref.[15]) (c) A scanning electron micrograph of a channel cross-section. Adapted from reference [16] and reproduced with permission.

1.2.1 Electrically Driven Ion Transport

We have studied the electrically driven transport of ions in our nanofluidic channels.[17] The ionic current was measured while a DC voltage, ΔV, was applied across a channel filled with aqueous solution of a given potassium chloride (KCl) salt concentration, *n*. The salt dependence of the conductance is shown in Figure 1.2 for 5 channels ranging in height from $h = 70$ nm to $h = 1050$ nm. At high salt concentrations, the channel conductances scaled with the salt concentration and the channel height, just as would be expected for a bulk KCl solution. For low salt concentrations, however, the conductance saturated at a minimum value independent of the channel height, and was orders of magnitude higher than would be expected from the bulk conductivity of the fluid.

The ionic conductance saturation results from the electrostatic influence of the charged channel walls on the ionic fluid. The silica surface is negative in solution at neutral pH, and therefore attracts positive counter-ions, while repelling negative co-ions. The thin region of fluid near the surface in which a net charge density is created is called the double layer.[18] It is the transport of mobile counter-ions in the double layer that accounts for the extra conductance observed at low salt concentrations.

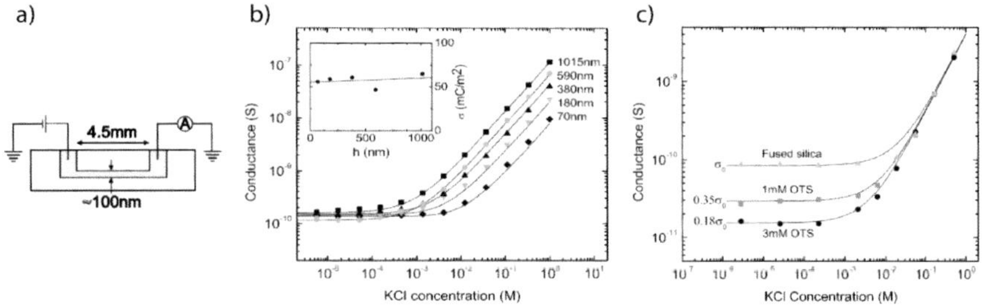

Figure 1.2 Surface-charge-governed ion transport in nanofluidic channels. (a) Cross-sectional illustration of a channel and the measurement apparatus configuration. (b) Salt concentration dependence of the DC ion conductance in a 50 μm wide channel. The solid lines are fits to the ion transport model described in the text. The values of σ obtained from the fits are plotted against h (inset). (c) The conductance of 87 nm high channels filled with 50% isopropanol, 50% KCl solution. The channels were treated with the indicated concentrations of OTS. Adapted from reference [17] and reproduced with permission.

The conductance of nanofluidic channels can be understood quantitatively. It is necessary to account for all the ions, including the double layer, and properly couple their motion to that of the fluid. We have modelled the electrostatic potential in the double layer using the nonlinear Poisson-Boltzmann (PB) equation, which is the conventional mean field theory that describes the competition between electrostatic and entropic forces on the ions:

$$\frac{d^2\psi(x)}{dx^2} = \kappa^2 \sinh(\psi(x)) \tag{1.1}$$

Here $k_BT\psi(x)/e$ is the electrostatic potential at height x from the channel mid-plane, e is the electron charge, k_BT is the thermal energy, $1/\kappa$ is the Debye screening length, defined by $\kappa^2 = 2e^2n/(\varepsilon\varepsilon_0 k_BT)$, and $\varepsilon\varepsilon_0$ is the permittivity of water. The Debye length sets the range of electrostatic interactions in solution. It is inversely related to salt concentration, increasing from $1/\kappa = 1$ nm at the roughly physiological salt concentration of $n = 100$ mM, to $1/\kappa = 10$ nm at $n = 1$ mM, and to $1/\kappa = 1$ μm in de-ionized water.

The exact solution for $\psi(x)$ in the slab geometry is known,[19] which allows us to calculate the exact (mean field) distribution of ions in our channels. The solution remains valid even when the double layers from opposing channel walls overlap. Moreover, the motion of ions is coupled to the fluid flow via the Stokes equation:

$$\eta\frac{d^2u(x)}{dx^2} - \frac{\Delta V}{l}\frac{\varepsilon\varepsilon_0 k_BT}{e}\frac{d^2\psi(x)}{dx^2} + \frac{\Delta p}{l} = 0 \tag{1.2}$$

where $u(x)$ is the fluid velocity, Δp is the pressure difference across the channel, and l is the length of the channel. We take $\psi(x)$ to be the equilibrium distribution, which is justified as long as the applied electric field gradients are too weak to significantly distort the double layer, i.e. smaller than $k_B T\kappa$.[20] It is also conventional to apply the no-slip boundary condition at the channel surfaces.

In the absence of an applied pressure gradient and taking the electrical mobility of the ions to be the bulk value, the solutions to Equations 1.1 and 1.2 can be used to calculate the total conductance of a channel. This was the approach used by Levine to calculate the ionic conductance in a narrow channel with charged walls.[21] However in order to accurately describe our experimental conductance data, it was necessary to replace the constant surface potential boundary condition that had been commonly used. We found that a constant effective surface charge density, σ, described the data extremely well and could be imposed on our transport model using Gauss' Law, i.e.

$$\sigma = \pm \frac{\varepsilon\varepsilon_0 k_B T}{e} \frac{d\psi}{dx}\bigg|_{x=\pm h/2} \tag{1.3}$$

Our ionic transport model described the experimental data very well, as can be seen from the theoretical fits in Figure 1.2(b). The model contains only a single fit parameter, namely σ, which was found to agree well with published values for silica surfaces obtained by chemical titration experiments.[22]

The ion transport model also provides insight into the very different behaviour that was observed in the high and the low salt regimes. At high n, the number of ions in the double layer is overwhelmed by the number in the bulk fluid. The conductance of a nanochannel at high n therefore increases with n just as the conductivity of bulk solution. At low n, by contrast, the counter-ions in the double layer dominate. Their number is fixed by the requirement of overall charge neutrality, and so the conductance of the nanochannel becomes governed by the charge density at the surfaces. The crossover between high-salt and low-salt behaviour occurs when $|\sigma| \approx enh$ for monovalent salt. It is important to note that this does not correspond to double layer overlap. The data in Figure 1.2(b) clearly show, for example, that a 380 nm high channel is in the low-salt conductance plateau at $n = 10^{-4}\,\mathrm{M}$, where the Debye length is only 30 nm.

Solid-state nanopores and nanotubes are systems in which ion transport in the low salt regime is particularly relevant. Due to their small diameter (<10 nm typically), the onset of the conductance plateau in a nanopore occurs at salt concentrations as high as hundreds of millimolar. In addition, nanopore experiments typically involve the insertion of an individual DNA molecule, which is itself a highly charged object. The backbone of double-stranded DNA carries two electronic charges for every 3.4 Å of length. DNA insertion into a solid-state nanopore therefore entrains a high concentration of mobile counter-ions into the pore, which actually increases the measured conductance for salt concentrations below ~0.4 mM.[23-25]

The electrically driven transport of ions in nanochannels reveals an interesting parallel with integrated circuits. The dependence of channel conductance on the surface charge is analogous to the conductance modulations in a field effect transistor (FET) that can be induced by the charge on the gate. It is therefore possible to "gate" the conductance of a nanofluidic channel by chemically modifying its surface charge density, as we have shown in Figure 1.2(c). The conductance of an h = 87 nm channel in the low-salt regime was clearly reduced by treatments with octadecyltrichlorosilane (OTS), whose attachment to silica neutralizes the surface. Other groups have employed this phenomenon as a sensing mechanism for biological agents[26] or reported how the surface charge density of a nanochannel can be voltage-modulated using gate electrodes to result in an "ionic transistor".[27,28]

1.2.2 Streaming Currents

Ions are displaced in a pressure-driven flow because of the viscous drag between them and the fluid. In bulk solution, equal densities of positive and negative ions leave the fluid neutral, so no net charge transport occurs. In the vicinity of a charged surface, however, the excess of counterions in the double layer is advected by the flow and carries an electrical current. These so-called streaming currents can become increasingly important in nanofluidic channels, whose surface to volume ratio is particularly high.

We have measured streaming currents in nanofluidic channels between $h = 70$ nm and $h = 1147$ nm.[14] The relationship between the streaming current, I_{str}, and Δp was found to be linear, so we characterized a channel by its streaming conductance, S_{str}, defined as the slope $I_{str}/\Delta p$. The salt concentration dependence of S_{str} for a typical $h = 140$ nm channel is presented in Figure 1.3 and shows an extended plateau at low n that drops to a small fraction of the plateau value as n is increased beyond ~1 mM.

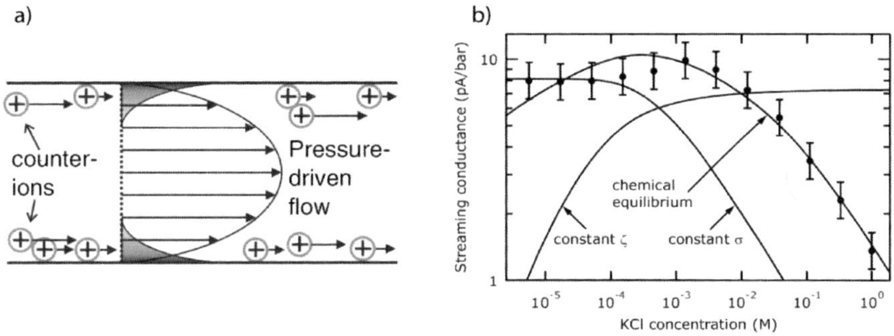

Figure 1.3 Streaming currents in nanochannels. (a) Schematic illustration of the origin of the streaming current. (b) Streaming conductance as a function of KCl concentration in a 140 nm high channel. The solid lines show model predictions for a constant surface charge, a constant surface potential, and a chemical equilibrium model discussed in the text. Adapted from reference [14] and reproduced with permission.

Streaming currents can be analyzed within the same theoretical framework as electrophoretic ion transport. The applied pressure, Δp, generates a parabolic (Poiseuille) fluid velocity profile that is maximal in the centre of the channel and stationary at the surfaces according to Equation 1.2. The distribution of ions that is described by the PB equation (Equation 1.1) is advected at the local fluid velocity. The streaming conductance is therefore highest at low n because the Debye length extends into the centre of the channel, where the fluid velocity is highest. We have found, however, that the constant σ boundary condition underestimates the streaming conductance at high n, predicting an earlier decay in S_{str} than observed. This can be resolved by accounting for the chemistry of the silica channel, whose surface charge density is taken to be salt and pH dependent using a model described by Behrens and Grier.[29] It predicts that as n increases, the double layer consists increasingly of potassium counter-ions rather than H^+. This shifts the chemical equilibrium towards a more negatively charged surface and explains the extended streaming current plateau that is observed in Figure 1.3(b).

At this point, we note the discrepancy between the boundary conditions that best describe pressure-driven and electrophoretic transport of ions in the same fluidic channels. This observation is not new. The Poisson-Boltzmann model can be used to interpret measurements of an object's charge by different techniques, including electrokinetic effects such as ionic conductance and streaming currents, as well as direct measurements of electrostatic forces on micron-scale surfaces using surface force apparatus[30] and atomic force microscopy (AFM) techniques.[31] It has been experimentally found that these techniques yield values for σ that can differ by a factor of 10 or more. These discrepancies highlight the fact that the Poisson-Boltzmann model does not accurately describe the microscopic structure of the double layer all the way down to the charged surface. As a result, it is necessary to speak of the "effective charge", which is a model-dependent parameter that characterizes a system's behaviour for a particular type of experiment. The double layer picture has been gradually refined to achieve more consistent predictions that account for the effects of non-specific adsorption (the so-called "Stern Layer"), ion correlations, and finite ion size.[32-34]

1.2.3 Streaming Currents as a Probe of Charge Inversion

Streaming currents are a sensitive probe of the surface charge and can be used to study the details of the solid-liquid interface.[35] The current derives from charge transport in the diffuse part of double layer only, because ions in the bulk fluid carry no net charge, and it is generally accepted that the tightly bound counter-ions in the Stern layer (also called the inner Helmholtz plane) remain immobile in a pressure-driven flow.[36-38] An important advantage of an electrokinetic probe of the surface charge over direct AFM force measurements is that streaming currents remain reliable even at high salt concentrations. We have used streaming currents in silica nanochannels to investigate the phenomenon of charge inversion (CI) by multivalent ions.

Ions play a fundamental role in screening electrostatic interactions in liquids. Multivalent ions (where the ion valency Z exceeds 1) can exhibit counterintuitive behaviour by not only reducing the effective charge of a surface, but by actually flipping its sign (Figure 1.4(a-b)). This phenomenon has been proposed to be relevant in important biological situations such as DNA condensation, viral packaging, and drug delivery.[39-41] CI, however, cannot be explained by conventional mean-field theories of screening such as the Poisson-Boltzmann model.

Shklovskii proposed an analytical model that assumes that multivalent counter-ions form a two-dimensional strongly correlated liquid (SCL) at charged surfaces and invert the surface charge above a critical concentration.[42] This effect is driven by the interaction parameter $\Gamma = \frac{\sqrt{|\sigma_b Z^3| e^3/\pi}}{4\varepsilon\varepsilon_0 k_B T}$ and is therefore strong for high Z and bare surface charge, σ_b. Besteman used AFM force measurements[43,44] to show that the SCL model accurately describes the dependence of c_0 on surface charge, dielectric constant, and ion valence for Z=3 and 4.

We first validated streaming current measurements as a new technique for studying CI by reproducing the findings for the Z=3 cation, cobalt(III)sepulchrate (CoSep), which is well understood both theoretically and experimentally. We then used streaming currents to test CI by divalent cations, for which earlier results had been inconclusive.[45-49] Figure 1.4(c) shows clear evidence for CI by the divalent cations Mg^{2+} and Ca^{2+} near the same high concentration of ~400 mM. This result is interesting for two reasons. First, these divalent ions are relevant in biology. Second, this is an example of CI in a regime where the inter-ionic coupling is weak. The validity of the SCL model does not extend to *Z*=2, so these experiments provide a practical guide to theoretical refinements.

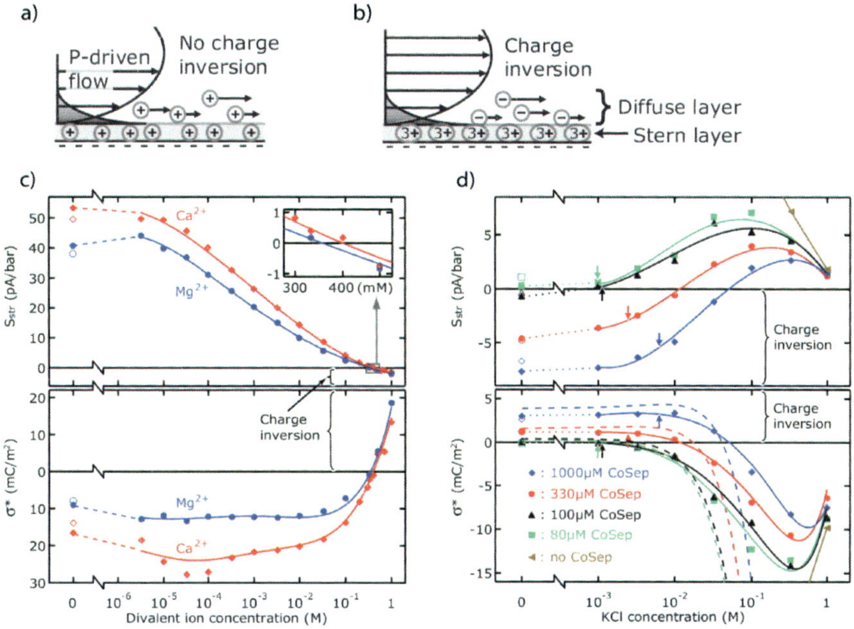

Figure 1.4 Studies of charge inversion using streaming currents. (a) Schematic illustration of the streaming current with screening by monovalent salt and (b) the effect of charge inversion by trivalent cations. (c) Divalent ion concentration dependence of (top) the streaming conductance and (bottom) the effective surface charge. The solid lines are guides to the eye; open symbols indicate measurements after each sweep from low to high concentration. The inset highlights the charge inversion concentration region. (d) The streaming conductance (top) and the effective surface charge (bottom) are plotted as a function of KCl concentration for various CoSep concentrations. The solid lines are guides to the eye; the dashed lines are model curves discussed elsewhere. Adapted from reference [35] and reproduced with permission.

The streaming current technique also allowed us to investigate the effects of monovalent salt on CI. High concentrations of monovalent salt (~150 mM) are typically present under physiological conditions in biological systems. This is expected to lead to screening of the surface charge and of multivalent ions. It was unclear, however, how this would affect CI. Our measurements showed that increasing concentrations of monovalent salt weaken and ultimately cancel CI by the trivalent cation CoSep (Figure 1.4(c)). The influence of monovalent salt on CI could be understood within a refined model of the SCL model up to the moderate concentrations at which CI was negated.

1.2.4 Electrokinetic Energy Conversion in Nanofluidic Channels

Electrokinetic phenomena exhibit a coupling between the transport of fluid and electricity. This presents interesting technological opportunities, such as for the electrical pumping of fluid by electro-osmosis, which has become important in microfluidics. In a reciprocal fashion, a pressure-driven fluid flow through a narrow channel carries a net charge with it that induces both a current and a potential when the charge accumulates at the channel ends. These streaming currents and streaming potentials can drive an external load, and therefore represent a means of converting mechanical work into useful electrical power. The notion of employing electrokinetic effects in an energy conversion device is not new,[50] but has received renewed attention in the context of micro- and nanofluidic devices, whose geometries and material properties can be engineered to optimize performance.[16,51-54] High energy-conversion efficiency and high output power are the requirements for such a device to be practical. We have evaluated the prospects for electrokinetic energy conversion both theoretically[16] and experimentally.[54]

A fluidic device capable of electrokinetic energy conversion consists of an inlet and an outlet that are connected by one or more channels with charged walls (Figure 1.5). Its electrokinetic properties in the linear regime are defined by the response of the ionic current, I, and the volume flow rate, Q, to the application of an electrochemical potential difference, ΔV, or a pressure difference, Δp, between the inlet and the outlet according to:

$$I = \frac{dI}{d\Delta p}\Delta p + \frac{dI}{d\Delta V}\Delta V = S_{str}\Delta p + \frac{\Delta V}{R_{ch}} \tag{1.4}$$

$$Q = \frac{dQ}{d\Delta p}\Delta p + \frac{dQ}{d\Delta V}\Delta V = \frac{\Delta p}{Z_{ch}} + S_{str}\Delta V \tag{1.5}$$

where R_{ch} is the electrical resistance, Z_{ch} is the fluidic impedance, and S_{str} is the streaming conductance of the channel. Only three quantities are needed to fully describe the device because the voltage-driven fluid flow rate and the pressure-driven ionic current are related by the Onsager identity,[55] $dQ/d\Delta V = dI/d\Delta p \equiv S_{str}$ which expresses the reciprocity between electrically induced fluid flows and flow-induced electrical currents.

The energy conversion efficiency is defined as the electrical power consumed by the external load divided by the input mechanical pumping power and is found to have a maximum value of $\varepsilon_{max} = \alpha \big/ \left(\alpha + 2\left(\sqrt{1-\alpha} + 1 - \alpha\right)\right)$ at the optimized load resistance $R_L = R_{ch} \big/ \sqrt{1-\alpha}$, where $\alpha = S_{str}^2 Z_{ch} R_{ch}$. The values of R_{ch}, Z_{ch}, and S_{str} for a particular device are determined by its geometry and the distribution of ions and fluid flows within it. We have used the theoretical framework outlined above to theoretically evaluate ε_{max} for slit-like channels.

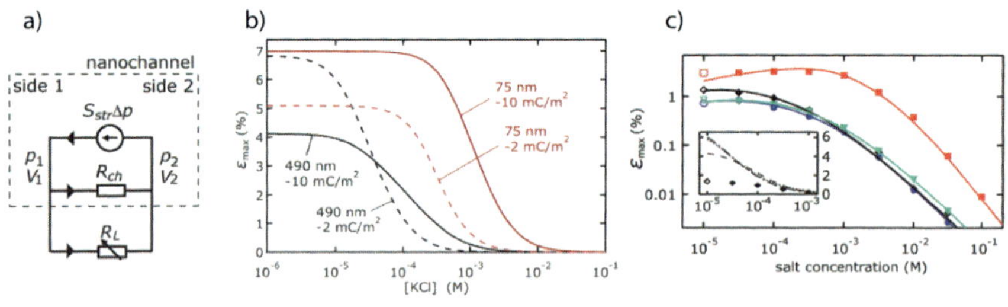

Figure 1.5 The efficiency of electrokinetic energy conversion in a nanofluidic channel. (a) Equivalent circuit of a nanochannel connected to a load resistor. (b) KCl concentration dependence of ε_{max} calculated using the model described in the text. The channel height and surface charge are as indicated. (c) Measured salt concentration dependence of ε_{max} for KCl, h = 75 nm (red); KCl, h = 490 nm (red); and KCl (blue) and LiCl (green) for the same h = 490 nm channel. Adapted from references [16, 54] and reproduced with permission.

The calculated n-dependence of ε_{max} shows that the highest energy conversion efficiency occurs in a low-*n* plateau, followed by a decay towards zero efficiency at high *n* (Figure 1.5(b)). The plateau region extends to higher *n* for smaller channels: up to $n = 10^{-5}$ M for a $h = 490$ nm channel and up to $n = 10^{-4}$ M for a $h = 75$ nm channel. This behaviour can be understood intuitively; in thin nanochannels and at sufficiently low salt concentrations, the double layers of opposing channel surfaces overlap, and electrostatic forces expel co-ions from the channel leaving only counter-ions. Because co-ions do not contribute to the electrical power generated by streaming currents, but instead provide an additional pathway for power dissipation through ionic conductance, they can only detract from the energy conversion efficiency. In addition, the extended double layers increase the concentration of counter-ions in the centre of the channel, where the fluid velocity is highest.

Our experiments confirmed that the maximum energy conversion efficiency occurs in the low salt regime. Figure 1.5(c) shows our experimental study of ε_{max} as a function of salt concentration for three channel heights. The energy conversion efficiency was found to be roughly constant at low *n*, and then decreased strongly at higher *n*. As predicted, the transition between the low and high salt regimes occurred at higher salt concentrations for smaller channels, for example, at $10^{-3.5}$ M for $h = 75$ nm versus $10^{-4.5}$ M for $h = 490$ nm.

We measured a peak energy conversion efficiency of ~3% for $h = 75$nm and KCl solution. This was less than half of the 7% efficiency that was predicted. We noted that this discrepancy could be reconciled by positing a finite conductance in the Stern layer, which is taken to be a layer of mobile counter-ions behind the no-slip plane on the channel. The counter-ions in the Stern layer consequently dissipate energy by electrical conductance without being advected by the pressure-driven flow. This model of the double layer was used to fit R_{ch}, S_{str}, and the efficiency data in Figure 1.5(c).

Clearly, a strategy is needed to significantly enhance efficiency if electrokinetic energy conversion is to be practical. An intriguing possibility that has recently been considered is to induce hydrodynamic slip at the surface of a channel.[56-58] A finite fluid velocity at the channel surface would increase the transport of counter-ions that concentrate there and eliminate the Stern layer by placing the extrapolated no-slip plane behind the channel surface. Recent experimental work and molecular dynamics simulations point to smooth and hydrophobic surfaces as conditions that promote slip.[59] The implication of a moderate degree of slip, characterized by a 30 nm slip length, is an energy conversion efficiency predicted to be 40%.[58] The extraordinarily long slip lengths recently reported for carbon nanotubes[60,61] imply very useful efficiencies exceeding 70%.

1.3 POLYMER TRANSPORT

Polymers are fundamental to biology. The genetic programs of all living systems are stored, translated, and executed at the molecular level by specialized polymers – DNA, RNA, and proteins, respectively.[62] These molecules contain valuable information regarding identity, biological function, and disease. This makes polymers important targets of lab-on-a-chip bioanalysis.

Length-separation is the most widespread analytical task for polymers, and is commonly accomplished by electrophoresis in a sieving medium such as a gel.[63,64] The interactions between polymers and the gel lead to length-dependent migration velocities.

Lab-on-a-chip devices were first reduced in size from the microscale to the nanoscale in a quest to control such interactions and create artificial separation devices. In nanofluidic channels small enough to restrict a polymer's internal degrees of freedom, it was found that entropic forces become important, giving rise to effects that have been exploited in novel separation strategies such as artificial gels[3] and entropic trap arrays.[65]

The transport of polymers within microfluidic and nanofluidic channels remains of central importance to lab-on-a-chip technology, but our understanding of the topic is far from complete. Polymers can be subjected to a wide variety of confining geometries, fluid flows, or electric fields. In this section we summarize our efforts to understand how polymers behave in situations that commonly arise in nanofluidics, such as parabolic fluid flows and electric fields.

1.3.1 Pressure-Driven Polymer Transport

It is straightforward to apply a pressure difference across a fluidic channel, and this has long been used to drive transport in chromatographic chemical separations. In contrast with electrical transport mechanisms such as electrophoresis or electro-osmotic flow, a pressure gradient generates a parabolic flow profile that leads to hydrodynamic dispersion, called Taylor dispersion, and flow speeds that depend strongly on the channel size.[10] These perceived disadvantages have made pressure-driven flows less popular than electrokinetic mechanisms in micro- and nanofluidic applications. The interaction of a parabolic flow with a confined flexible polymer had consequently remained unexplored.

We investigated the mobility and dispersion of long DNA polymers in a pressure-driven fluid flow in slit-like nanofluidic channels.[15] The centre-of-mass motion of individual molecules was tracked by epifluorescence optical microscopy and analyzed to determine the mobility and the dispersion of DNA. We investigated the influence of applied pressure, channel height and DNA length, L. Our results reveal how length-dependent and length-independent transport regimes arise from the statistical properties of polymer coils, and that the dispersion of polymers is suppressed by confinement.

1.3.1.1 Pressure-Driven DNA Mobility

Our experiments showed that the mean velocity of DNA, $\overline{V}$, increased linearly with Δp for a given L and h. To best reveal the dependence of pressure-driven DNA transport on L and h, we first defined a pressure-driven DNA mobility, ν, as the slope of $\overline{V}$ vs. Δp, i.e. $\overline{V} = \nu\Delta p$. We then plotted the ratio ν/ν_λ for each DNA length in the same channel, where ν_λ is the mobility of 48.5 kbp-long λ-DNA (Figure 1.6). This approach provided a measure of mobility that was insensitive to microscopic channel irregularities.

Two distinct regimes of pressure-driven transport can clearly be identified in Figure 1.6. In large channels ($h > \sim 2$ μm), the mobility of DNA increased with polymer length, with the mobility of λ-DNA exceeding that of 8.8 kbp-long DNA by 12% in h = 3.8 μm channels. In thin channels ($h < \sim 1$ μm), the mobility was found to be independent of length within experimental error. In order to explain the pressure-driven motion of DNA, we must consider the statistical distribution of DNA across a slit-like nanochannel. Thermal forces contort a DNA polymer into conformations that simultaneously sample different regions of the fluid flow. We assume that each segment of DNA moves at the local fluid velocity and take the centre-of-mass velocity to be the average over all segments, according to:

$$\bar{V} = \int_{-h/2}^{h/2} \rho(x)u(x)dx \Big/ \int_{-h/2}^{h/2} \rho(x)dx \tag{1.6}$$

where $\rho(x)$ is the average concentration of DNA segments across the channel. We can calculate $\rho(x)$ by modelling the polymer coil as a random flight whose equilibrium conformations are described by the Edwards diffusion equation[66]:

$$\frac{b^2}{6}\nabla^2 P(x,s) = b\frac{\partial P(x,s)}{\partial s} \tag{1.7}$$

where $P(x,s)$ is the probability that paths of contour length s end at height x, and b is the Kuhn length, which characterizes the stiffness of the polymer. From this probability distribution, the average density of polymer segments is given by,

$$\rho(x) = \int_0^L P(x,s)P(x,L-s)dx \tag{1.8}$$

The confinement of a polymer to a slit geometry was first treated by Casassa[67] and Casassa and Tagami[68] by imposing non-interacting boundary conditions at the walls, i.e. $P(\pm h/2) = 0$. We have used Casassa's exact result for $P(x)$ to calculate the mobility of DNA polymers in our channels. The values of b were fixed by matching the known radius of gyration for each polymer length, R_g, with the coil size of a random flight polymer according to $R_g = \sqrt{(Lb)/6}$. We note that this procedure implicitly includes polymer self-exclusion effects, which are not explicitly modelled by Equation 1.7.

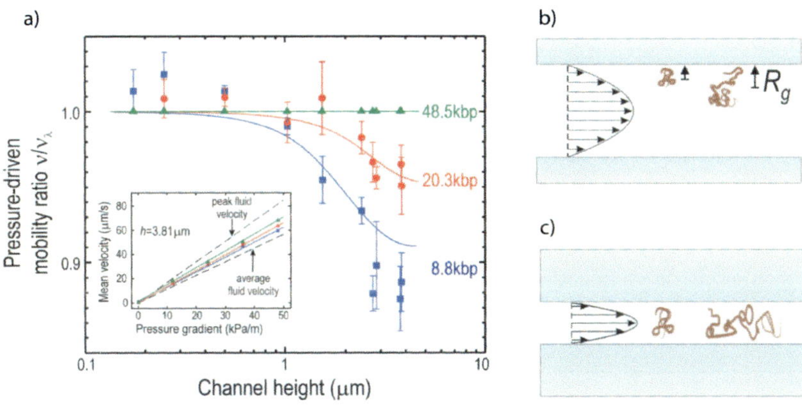

Figure 1.6 Pressure-driven DNA mobility in nanofluidic channels. (a) (inset) The average velocity of DNA molecules in an h = 2.73 μm channel versus applied pressure gradient. The slope of the curve defines the pressure-driven mobility, ν. (main panel) The ratios of ν for the indicated DNA fragment lengths to ν_λ for 48.5 kbp-long DNA. The solid lines indicate predictions of the polymer transport model described in the text. (b) Schematic DNA configurations in a wide channel, where the molecules' centre of mass is excluded from a region of length R_g from the channel wall. Long molecules are confined to the high velocity region of the flow, and consequently move faster than shorter molecules. (c) Schematic DNA configurations in a narrow channel, where DNA mobility is independent of length. Adapted from reference [15] and reproduced with permission.

The results of our polymer transport model are plotted together with the mobility data in Figure 1.6. The model clearly predicts both length-dependent and length-independent regimes, as well as the crossover between them for all DNA lengths. The quantitative agreement between our model and the data is also good. The model only mildly overestimates the mobility of the shortest DNA fragment in the largest channels. We note that our model contains no fitting parameters. Its only inputs are the known lengths and radii of gyration of the DNA fragments.

The physical picture behind the two pressure-driven transport regimes can be understood as follows: In channels that are large compared with the polymer coil size, molecules can diffuse freely in the central region of the channel. However, they are impeded from approaching the walls closer than the radius of gyration. Long DNA molecules are therefore more strongly confined to the centre of the channel, where the fluid velocity is highest. This explains why long molecules travel faster in large channels and corresponds to the concept of "hydrodynamic chromatography" of polymers.[69] In channels that are small compared with the polymer coil size, the DNA is squeezed into new conformations. Although the shapes of longer molecules may be more extended in the plane of the channel, the concentration profile of DNA segments across the channel height is length-independent. This explains why the mobilities of long and short DNA molecules are the same in the thin channel regime.

1.3.1.2 Dispersion of DNA Polymers in a Pressure-Driven Flow

The dispersion of DNA arises from the fluctuations in a molecule's velocity. We have used the same pressure-driven DNA measurements to study this fundamental transport property. We quantified velocity fluctuations by the dispersion coefficient, D^*, defined by $2D^*\Delta t = \langle(\Delta z - \overline{V}\Delta t)^2\rangle$. Here $\langle(\Delta z - \overline{V}\Delta t)^2\rangle$ is the mean square displacement of a molecule from its original position along the channel length, translated with the mean velocity in the time interval Δt.

Dispersion in a pressure-driven flow is understood to originate from two distinct mechanisms. First, the random thermal forces exerted on a particle by the surrounding fluid give rise to Brownian motion, which we parameterize by the thermal self-diffusion coefficient, D_0. Second, hydrodynamic dispersion also arises in a non-uniform flow. A particle in a fast moving region of the flow is pulled ahead of a particle near the wall, where the flow is slow. Taylor first considered the interplay between these thermal and hydrodynamic dispersion mechanisms for point particles, and so the phenomenon is known as Taylor dispersion.[70]

Generalized Taylor dispersion theory[71] predicts that D^* can be expressed as the sum of a purely thermal term and a convective term proportional to $\overline{V}^2$:

$$D^* = D_0 + \alpha_T \overline{V}^2 \tag{1.9}$$

Here, we have introduced a parameter, α_T, to quantify the hydrodynamic component of dispersion. We have found that Equation 1.9 fits our experimental data extremely well, as shown in Figure 1.7(a) for two different channel heights. We have used such fits to extract experimental values for α_T.

The thermal self-diffusion of DNA was studied by considering dispersion in the absence of an applied pressure (where $\overline{V} = \alpha_T = 0$). We found that D_0 was suppressed with increasing confinement in thin channels, in good agreement with a famous scaling relationship predicted by de Gennes.[72] This confirmed previous observations of the phenomenon[73-75] and extended its verified range to higher degrees of confinement. More recent measurements by the Doyle group have found subtle departures from the de Gennes predictions.[76]

To date, the hydrodynamic dispersion of long polymers has not been realistically modelled, so an appropriate prediction of α_T is not available. We instead compared the experimentally determined values of α_T to models of point-particle[77] and free-draining rigid sphere dispersion[69] as shown in Figure 1.7(b). Both models clearly fail to describe the observed behaviour. Taking DNA to be a point particle overestimates its dispersivity by more than an order of magnitude, whereas the rigid sphere model predicts a rapid drop to zero dispersion that was not observed.

These results suggest that Taylor dispersion theory needs to be extended in order to account for the unusual behaviour of polymers under confinement. Our experiments have nonetheless demonstrated empirically that the hydrodynamic dispersion of DNA polymers is greatly suppressed in nanochannels, which may be of benefit to analysis applications.

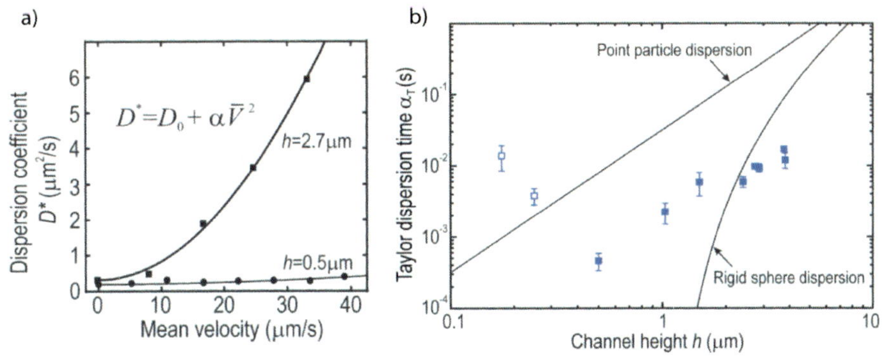

Figure 1.7 Taylor dispersion of DNA polymers in micro- and nanofluidic channels. (a) The dispersion coefficient of λ-DNA is plotted against its mean velocity for h = 2.73 and h = 500 nm. Solid lines indicate fits of $D^*=D_o+\alpha_T V^2$. (b) The height dependence of the measured values of α_T for 8.8 kbp-long DNA is compared with point-like and free-draining rigid particle models of Taylor dispersion.

1.3.2 Electrokinetic DNA Concentration in Nanofluidic Channels

For a chemical separation technique to be successful, an ensemble of sample molecules must first be collected in a narrow band from which they are typically "launched" into the separation region. The controlled concentration of analyte has therefore become an important goal of recent lab-on-a-chip research. Leading strategies include the use of nanofluidic filters,[78] micellar electrophoretic sweeping,[79] field-amplified sample stacking,[80] isotachophoresis,[81] electrokinetic trapping[82] and membrane pre-concentration.[83,84] We have recently discovered that in the nanofluidic regime, it is possible to take advantage of the unique transport properties of long polymers to concentrate them in an elegant way.[85] Our idea for DNA concentration was inspired by observations of surprising electrically driven DNA flow patterns in our slit-like nanochannels. Under a constant applied potential, DNA molecules aggregated near the end of the channel at the negative pole. At the same time, DNA was depleted from the opposite end of the channel. These dynamics were surprising because the axial velocity of DNA actually flips sign twice along the length of the channel, and also because this behaviour was only observed at low (<mM) salt concentrations. At high salt concentrations (~60 mM), DNA migrated continuously from one pole to the other.

We sought to understand the origin of these dramatic DNA flows at low n so that we might induce the controlled pre-concentration of DNA at arbitrary salt concentrations. A polymer transport model similar to the one outlined in Section 1.3.1.1 explained our observations. In the case of electrically driven transport, we considered two driving terms. First, electrophoresis moves DNA relative to the fluid with a migration velocity, $\vec{v}_{elec}$, that is proportional to the local electric field. Second, the fluid itself is driven by electric fields because counter-ions in the double layer drag the fluid by electro-osmosis. The fluid flow then carries the DNA by advection with a velocity $\vec{v}_{advect}$.

A highly symmetric situation that is commonly found in micro- and nanofluidics is a constant electric field in a uniform channel. In this case $\vec{v}_{elec}$ and $\vec{v}_{advect}$ are both constant and proportional to each other. In general, however, the electric field can vary along the length of the channel, as can the electrophoretic force at the channel walls, e.g. due to conductivity and surface charge density variations, respectively. This establishes a competition between electrophoresis and advection that is well illustrated by the following example (Figure 1.8).

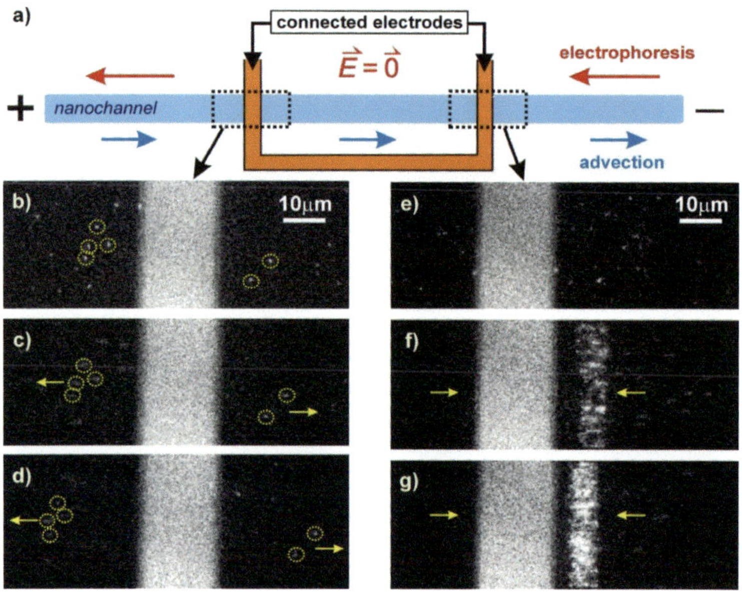

Figure 1.8 Inducing controlled DNA concentration and depletion in a 500 nm high nanofluidic channel. (a) Schematic top view of the device, showing the fluidic nanochannel intersected at two points by the connected electrodes. The electrodes do not block the channel but contact the fluid in order to cancel the electric field between them. The red and blue arrows indicate the predicted $\vec{v}_{elec}$ and $\vec{v}_{advect}$ in the different channel regions, respectively. Fluorescence micrographs show DNA molecules near the electrode towards the positive pole prior (b) to the application of 10V across the channel and then at increasing times after (c and d). The same molecules are circled in each image to highlight their motion away from the electrode. Near the electrode towards the negative pole, DNA molecules are shown prior to (e) the application of 10V and then after (f)–(g). Adapted from reference [85] and reproduced with permission.

We designed a nanofluidic device to concentrate DNA at a specified location, and at technologically relevant salt concentrations. The layout consisted of a slit-like channel intersecting an annular gold electrode at two points. The electrode was fabricated so as to contact the fluid in the channel, and maintain a constant electrochemical potential between the two intersection points. The electrode was embedded into the bottom surface so as to maintain the channel cross-section, and avoid impeding the fluid flow.

When a constant ΔV was applied across the channel, strong electric fields were generated near the ends, so $\vec{v}_{elec}$ dominated $\vec{v}_{advect}$ there, and DNA was pulled towards the positive pole. In the region between the electrodes, $\vec{v}_{elec}$ was suppressed. $\vec{v}_{advect}$, on the other hand, continued to push DNA towards the positive pole in the central region because the fluid is effectively incompressible and travels towards the negative pole at a constant flow rate Q. The net effect of these driving forces was to concentrate DNA at the electrode near the negative pole and deplete it from the other electrode. We note here that the details of the induced fluid flow are in fact complicated, involving re-circulating motion that would disperse point particles distributed across the channel height. It is the ability of a confined polymer to sample the entire flow profile that leads to an averaged value for $\vec{v}_{advect}$ and the possibility to electrically concentrate DNA. This nanofluidic pre-concentration method for polymers is appealing because it is entirely electrically driven. It is therefore straightforward to implement and control, and the integrated electrodes can also be used for other functions, such as driving the molecules into a separation device.

1.3.3 DNA Conformations and Dynamics in Slit-Like Nanochannels

In studying the transport of DNA polymers in nanofluidic channels, we have mainly focused on how polymers interact with imposed electric fields and fluid flows. The specific conformations that a molecule adopts, and their characteristic fluctuation rate, play important roles as the molecule interacts with its environment, e.g. the features of a separation device. We have recently sought to better understand the static and dynamic properties of DNA in confined environments, which is important for the design of single-molecule analysis and manipulation devices, and may provide insight into natural processes like DNA packaging in viruses[86] and DNA segregation in bacteria.[87] Here we outline some of our recent results on DNA conformations and dynamics in slit-like channels.[88]

We studied the conformations of DNA polymers under increasing degrees of confinement from their two-dimensional projections (Figure 1.9(a)). We analyzed fluorescent DNA images to compute the radius of gyration tensor for each conformation. The principal axes of these conformations were calculated from the eigenvalues of the radius of gyration tensor and used to quantify properties such as the coil size and molecular anisotropy. The measured distributions of the major and minor axis lengths are shown in Figure 1.9(b) (along with theoretical fits based on a statistical model that is presented elsewhere[88]). The dependence of the measured anisotropy on confinement was found to be consistent with the theoretical model by van Vliet and ten Brinke,[89] i.e. for decreasing h, the anisotropy first decreased due to an alignment of the longest molecular axis with the channel by rotation. Anisotropy then increased as confinement caused excluded volume interactions to stretch the molecule laterally.

Our experiments also found that the polymer coil size went through three distinct regimes with increasing confinement. The DNA was first compressed slightly in a manner consistent with predictions of van Vliet *et al.*[89] It then extended with confinement as predicted by the well-known scaling theory by de Gennes.[72] In the thinnest channels, whose height was comparable to the DNA persistence length, the DNA size reached a plateau. This last regime is often referred to as the "Odijk" regime.[90,91] The characteristic relaxation time of the molecule, given by the autocorrelation of its size fluctuations, also revealed the same three regimes. The relaxation time decreased slightly with h in the highest channels, followed by a rapid rise in the de Gennes regime, and then a rapid decline in the Odijk regime. These observations of DNA in slits are consistent with previous observations of de Gennes and Odijk scaling in square channels. We note that the Odijk regime is somewhat different in the case of wide slits, because the polymer can bend back on itself without having to pay a large energetic price.

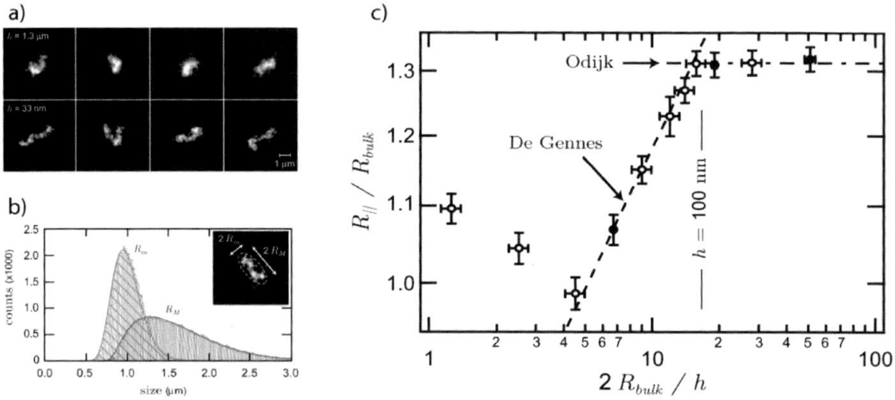

Figure 1.9 DNA conformations in nanofluidic slits. (a) Typical images of λ-DNA in 1.3 μm and 33 nm high channels. (b) Histograms of the major (R_M) and minor (R_m) axis lengths of λ-DNA for h = 107 nm. (inset) An image of a molecule indicating the calculated R_M and R_m. (c) The projection of the radius of gyration as viewed from above the slit versus confinement. The dashed line indicates the de Gennes scaling relationship. The dash-dotted line indicates the Odijk regime. Adapted from reference [88] and reproduced with permission.

1.4 MICROTUBULE TRANSPORT IN NANOFLUIDIC CHANNELS DRIVEN BY ELECTRIC FIELDS AND BY KINESIN BIOMOLECULAR MOTORS

When Richard Feynman contemplated the challenges associated with manipulating matter at the nanoscale back in 1959,[92] he could turn to biology for inspiration on how to cope with these issues. The biological cell contains the ultimate in nanomachines, with processes such as cell motility, energy production, protein assembly, cell division and DNA replication all originating from the activities of small protein machines.[93,94] Examples include the rotary motor that drives the bacterial flagella,[95] and the linearly moving motors, for instance, myosin that drives muscle contraction, or kinesin and dynein motors that drive the intracellular transport of materials.[96]

Biomotors are enzymes that contain moving parts and use a source of free energy to direct their mechanical motion. The linear-motion motors (kinesin, myosin, dynein) use the energy of the hydrolysis of adenosine-triphosphate (ATP), the cell's energy molecule, to move in discrete steps along tracks made of long protein polymers (actin filaments for myosin, microtubules for kinesin and dynein) that form the cytoskeleton that extends throughout the cell.[97] The structural polarity of these filaments (denoted by a plus and minus end) allows unidirectional movement of the motors along their tracks. For example, a single kinesin consists of two motor domains that can step in discrete 8.2 nm steps (corresponding to the periodicity of a microtubule's protofilament) toward the plus end along a microtubule.[98] Upon each 8.2 nm step, kinesin can withstand an opposing force of up to 6 pN,[98,99] thus performing work of approximately 50 pN·nm. Since there is a tight coupling between a single step and the hydrolysis of a single molecule of ATP[100-102] (liberating ~80 pN·nm of free energy at cellular conditions), kinesins can work at an impressive ~60% efficiency.

One particularly active field of research is the use of biomotors for actuation and transport of materials in artificial nanofabricated environments.[103] In the first part of this chapter we describe recent advances in the exploitation of kinesin motors for nanotechnology. In particular we discuss the use of electric forces to manipulate the direction of individual microtubules at junctions. Thereafter, we describe the use of nanofabricated structures and electric fields for biophysical studies on single microtubules.

1.4.1 Electrical Manipulation of Kinesin-Driven Microtubule Transport

A molecular motor such as kinesin can potentially be used as the workhorse in miniaturized analytical systems or nano-electromechanical systems.[103-105] For example, active transport by molecular motors could be used for purification of materials against a flow, for the concentration of molecules, or for transport in increasingly smaller capillaries. These motors can be employed in either one of two geometries (Figure 1.10(a) and (b)). In the so-called bead assay (Figure 1.10(a)), mimicking the biological situation, a motor-coated cargo is transported along cytoskeletal filaments that are adsorbed onto a substrate. In the alternative geometry (Figure 1.10(b)), which is mostly employed, cytoskeletal filaments are propelled by surface-bound motors. In this gliding-assay geometry, the microtubules act as nanoscale trucks while transporting an attached cargo.

A considerable current effort in the field of molecular motor-assisted technology is aimed at building a nanoscale transport system in which biomotors are used for the active concentration of analyte molecules that are present in otherwise undetectably low quantities.[106-108] A schematic of such an envisioned device is pictured in Figure 1.10(c). Microtubule transporters, functionalized with antibodies specific to the analyte molecule of interest, bind molecules from a sample and transport them toward a second region on a chip, thereby concentrating the analyte and facilitating detection. En route, sensing and sorting capabilities are necessary.

An important step toward this goal has been the integration of kinesin motor proteins inside microfluidic channels.[109] The use of enclosed channels offers a great advantage for the confinement of motility to predetermined pathways. In the open trench-like structures that were used previously,[110-115] a common problem was that the motor-propelled microtubules collided with the sidewalls of the structures and were pushed out of their tracks. Another advantage of enclosed channels is that strong electric fields can be locally employed to manipulate individual (negatively charged) microtubules, as opposed to the large-scale effects of electric fields applied in open structures as previously demonstrated.[116-118]

We first show that microtubule motility can be reconstituted inside enclosed fluidic channels. To demonstrate this, we have fabricated micron-sized microfluidic channels in glass substrates (Figure 1.11(a)). We use e-beam lithography and wet-etching to fabricate a network of open channels in fused-silica substrates. The width of the channels varies from several to tens of micrometers, and the depth of the channels is typically one micrometer or less. The channel structures were sealed with a second glass substrate (Figure 1.11(a) and (b)). The insides of the channels were coated with casein and kinesin motor proteins by flushing the protein-containing solutions through the channels. Then, upon adding a solution containing microtubules and ATP, microtubules bind to the surface-adsorbed kinesins and are subsequently propelled through the channels. The movement of microtubules is imaged by fluorescence microscopy (Figure 1.11(c)).

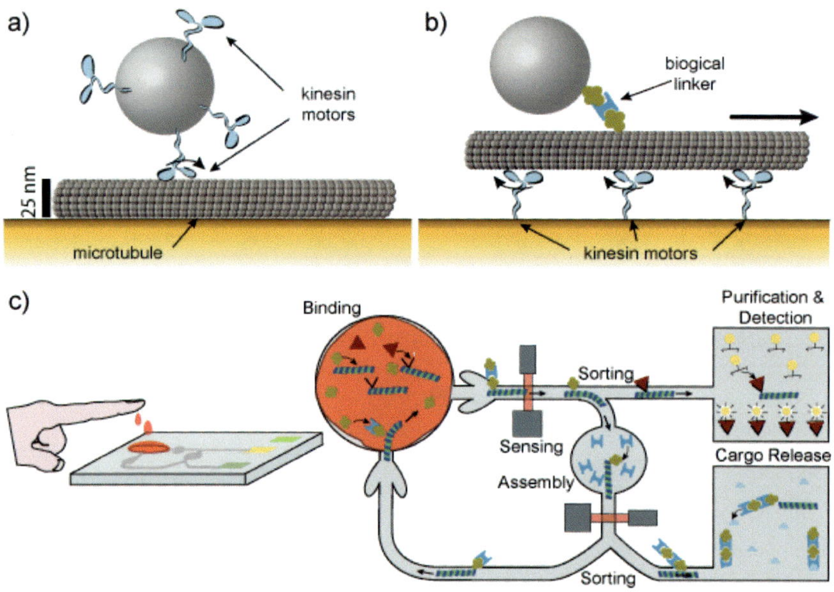

Figurel.10 Kinesin motors as transporters in artificial environments. (a) In the bead assay, motor-coated cargo moves along surface adsorbed cytoskeletal filaments. (b) In a gliding assay, surface-bound motors propel cytoskeletal filaments that can act as shuttles for a bound cargo. (c) Fictitious device combining diverse functionalities such as rectification and sorting of motility, purification and detection of analyte molecules, and the assembly and release of cargo molecules. Adapted from reference [103] and reproduced with permission.

We show that individual microtubules can be steered through application of electric fields. Steering of microtubules is a necessary prerequisite for imagined applications such as those depicted in Figure 1.10(c). To demonstrate this, we fabricated a Y-junction of channels across and through which a perpendicular channel was fabricated (Figure 1.11(d)). An electric field can be induced and confined inside this perpendicular channel through the application of a voltage difference between electrodes at either end of the channel. In Figure 1.11(d) we show a microtubule approaching a Y-junction, initially headed toward the left leg of the junction (t = 0 s). Upon application of an electric field in the perpendicular channel, an electric-field induced force acts in the opposite direction, gradually changing the course of the microtubule. Eventually it is steered into the right leg of the Y-junction (Figure 1.11(d), t = 20 s).

To illustrate the feasibility of a sorting application such as that depicted in Figure 1.10(c), we have employed electric forces to sort a population of microtubules carrying different cargos into different reservoirs on a chip (Figure 1.12). To this end we prepared and mixed microtubules that carry different colours (red and green) of fluorescent molecules. Using a colour-sensitive camera to discriminate between the microtubules, red microtubules approaching a Y-junction were actively steered into the left leg of the junction, whereas green microtubules were sent in the opposite direction, simply by reversing the polarity of the electric field (Figure 1.12(a)). After a large number of such successful single-microtubule redirections one reservoir contained predominantly red microtubules (91 %) and the other reservoir contained predominantly green microtubules (94 %) (Figure 1.12(b)).[109]

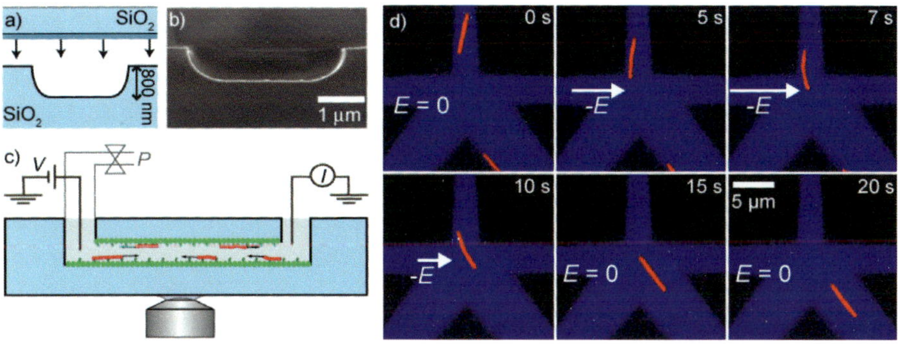

Figure 1.11 Electric forces can be used to steer individual microtubules in microfluidic channels. (a) Channels are fabricated in glass channels and sealed with a glass cover slip. (b) Scanning electron-microscopy image of a cross section of a channel. (c) Channels are coated with kinesin motor proteins and microtubules can be imaged moving through the channels by fluorescence microscopy. (d) A microtubule that approaches a Y junction from the top is steered through the use of electric forces into the right leg. The electric force points in the direction indicated, and the strength of the electric field is varied between 0 and 50 kV/m (indicated by length of arrow). Adapted from reference [109] and reproduced with permission.

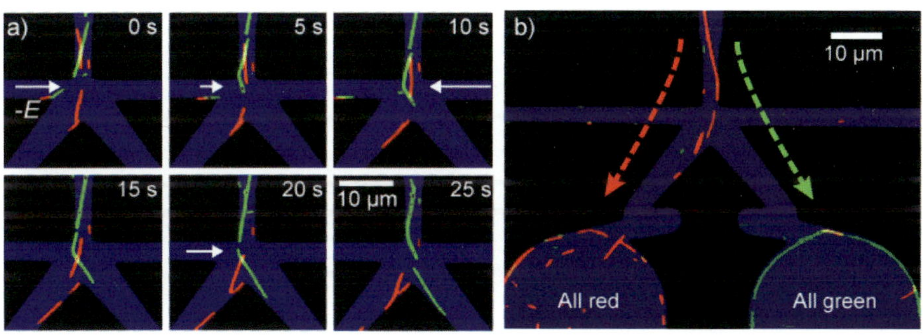

Figure1.12 On-chip sorting of a population of red and green-labeled microtubules. (a) By manually changing the polarity of the applied voltage, first a green microtubule is steered into the right leg of the junction (t = 0 and 5 s), then a red microtubule is steered into the left leg (t = 10 s), and finally a green microtubule is steered into the right leg again (t = 20 s). (b) The two legs of the junction lead to different reservoirs. After time, the right reservoir contains predominantly green microtubules, whereas the left reservoir contains red microtubules. Adapted from reference [109] and reproduced with permission.

In conclusion, the use of biomotors for sorting and transport of materials is an exciting development and could provide an interesting alternative to pressure-driven or electro-osmotic flow driven transport on a chip. However, the latter technologies are fairly well-developed, whereas many biomotor-powered applications are currently merely proof-of-principle demonstrations. At the moment, many speculative proposals for biomotor-driven applications still fail to be competitive when critically scrutinized and weighted against existing alternatives. Nevertheless, it may be that the application of biomotors for technology is currently only limited by the imagination and creativity of researchers and new opportunities may be found in yet unforeseen directions. Yet, the small size, force-exerting capabilities and possibilities for specific engineering of biomotors, for which currently no real alternatives exist, offers exciting opportunities that call for exploration.

1.4.2 Mechanical Properties of Microtubules Measured from Electric Field-Induced Bending

An interesting spin-off of the above mentioned experiments has resulted from the study of the mechanism that underlies the steering of individual microtubules. The curvature of microtubule trajectories under a perpendicular force of known magnitude is an amplification of the microscopic bending of their leading microtubule ends and thus provides an elegant measure of their stiffness. The stiffness of short microtubule ends, which are only sub-micrometer in our experiments, is of interest since recent experiments have indicated that the mechanical properties of microtubules on short length-scales can deviate considerably from their long-length behaviour.[119] Here, we describe experiments in which the observation of microtubule-trajectory curvatures under controlled perpendicular electric forces allows us to obtain an estimate of the mechanical properties of very short microtubule ends.[120] The use of nanofabricated structures for these experiments is beneficial, since it allows for a controlled and directed application of electric forces. Moreover, the high surface-to-volume ratio of microfluidic channels limits Joule-heating of the solution while an electric field is applied.

In order to study microtubule trajectories under perpendicular electric fields we fabricated channels in a perpendicular layout (Figure 1.13(a)). Microtubules enter the wide horizontal channel, in which a homogeneous electric field is present and, similarly to the steering experiments shown in Figure 1.11, their direction of motion is changed. In Figure 1.13(a) we show trajectories of microtubules that enter the electric field and gradually change their trajectory, in such a way to become aligned parallel and opposite to the electric field.

The mechanism of redirection is as follows (Figure 1.13(b)). The electric field induces a constant force density on the homogeneously charged microtubule in the direction opposite to the electric field. Kinesin molecules, distributed along the length of the microtubule, exert the opposing forces. This prevents movement of the microtubule perpendicular to its axis, which is confirmed by the data shown in Figure 1.13(b), where we show exactly overlapping traces of the leading and trailing-end coordinates of a microtubule moving in an electric field. Thus, the thermally fluctuating leading tip of the microtubule is biased into the direction of the applied force, thereby orienting the microtubule in a step-by-step fashion into the direction of the electric field.

The average curvature of the microtubule trajectory $\langle d\theta/ds \rangle$ (where the trajectory is described with coordinates θ and s (Figure 1.13(b)) relates to the perpendicular component of the applied field-induced force $f_\perp$ through the persistence length p and the average length $\langle d \rangle$ of the leading microtubule end as[120]

$$\left\langle \frac{d\theta}{ds} \right\rangle = \frac{\langle d \rangle^2}{3k_b T p} f_\perp \tag{1.10}$$

Equation 1.10 allows us to determine the value of the tip persistence length from observations of the curvature of microtubule trajectories under a known force provided that the average tip-length is known. The magnitude of the field-induced force $f_\perp$ is proportional to the electric field E that we apply,[120,121]

$$f_\perp = \mu_\perp c_\perp E \sin\theta \tag{1.11}$$

Here, $\mu_\perp$ is the mobility of a microtubule for motion perpendicular to its axis during free electrophoresis and $c_\perp$ is the perpendicular hydrodynamic drag coefficient per unit length of a microtubule close to a surface. The value of $\mu_\perp$ can be measured through electrophoresis experiments on individual microtubules,[122] as we show in the final section of this chapter, and amounts to $\mu_\perp = -(1.03 \pm 0.01)10^{-8}$ m^2/Vs for these experiments. The value of the perpendicular hydrodynamic drag coefficient of a microtubule was measured by Hunt *et al.* as $c_\perp = (1.19 \pm 0.11)10^{-2}$ Ns/m^2.[123]

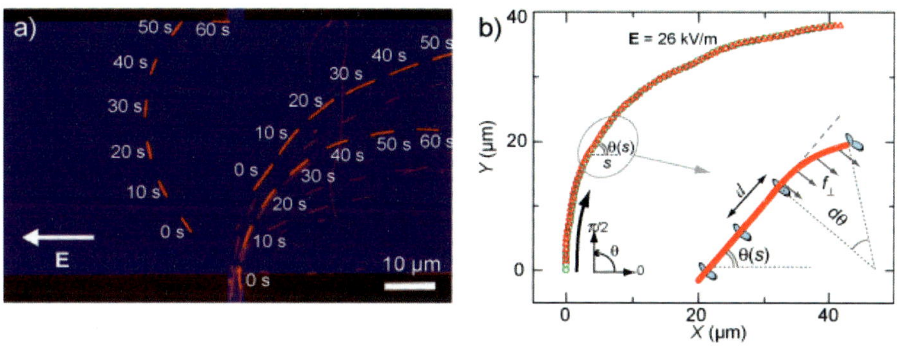

Figure1.13 The curvature of microtubule trajectories under an applied electric field. (a) Overlay of fluorescence images (with 10 s intervals) of microtubules moving under an electric field E = 26 kV/m. The microtubules enter from the small channel below and gradually become aligned with the electric field. (b) Overlapping coordinates of the leading (red triangles) and trailing (green circles) ends of a microtubule moving in a field E = 26 kV/m show that there is not motion of the microtubule perpendicular to its axis. The steering mechanism is thus due to the bending of the leading tip of the microtubule (inset). The change in orientation of the leading end of the microtubule (due to the force component perpendicular to the tip $f_\perp$), determines the trajectory curvature dθ/ds at any point s along the trajectory. Adapted from reference [120] and reproduced with permission.

Using fluorescence microscopy, we image trajectories of a large number of microtubules entering an electric field of magnitude E = 26 kV/m (i.e. under a field-induced perpendicular force $f_\perp$ = 3.2 ± 0.4 pN/μm for microtubules that have a 90° orientation with respect to the electric field (Equation 1.11)). We determine tangent trajectory angles θ and trajectory curvatures dθ/ds for all measured trajectories at every coordinate. Thus, we obtain a large number of orientation-invariant curvatures $d\theta/ds(\sin\theta)^{-1}$, the distribution of which we show Figure 1.14(a). We take the centre of this distribution as a measure of the mean-orientation invariant curvature at this particular force. We repeated these measurements for a range of electric fields between 0 and 44 kV/m. As expected (Equation 1.10), the mean orientation-invariant curvature increases linearly with E (Figure 1.14(b)). The red line is a linear fit through the data, and from the slope we determine that $\langle d\rangle^2/p = (1.30 \pm 0.16)10^{-10}$ m (Equations 1.10 and 1.11).

Finally, we obtain an estimate for the average tip length $\langle d\rangle$ from observations of the trajectories of very short microtubules that move without any applied electric field. Long microtubules are bound to and propelled by several kinesin molecules distributed along their length and will therefore preserve their directionality. However, if the microtubule length becomes small and comparable to $\langle d\rangle$, then occasionally the filament will be bound to only a single kinesin molecule and display diffusive rotational motion around the motor, thereby rapidly changing its orientation.[123,124] There is a clear relation between the average distance travelled by the microtubule between successive rotations and the ratio of the microtubule length L and the average tip length $\langle d\rangle$[124] Thus, a measurement of $\langle S\rangle$ for a microtubule of known length L provides a measurement of the average tip length $\langle d\rangle$. Using this method, we have measured that in our experiments the average tip length $\langle d\rangle = 0.10 \pm 0.02$ μm.[120]

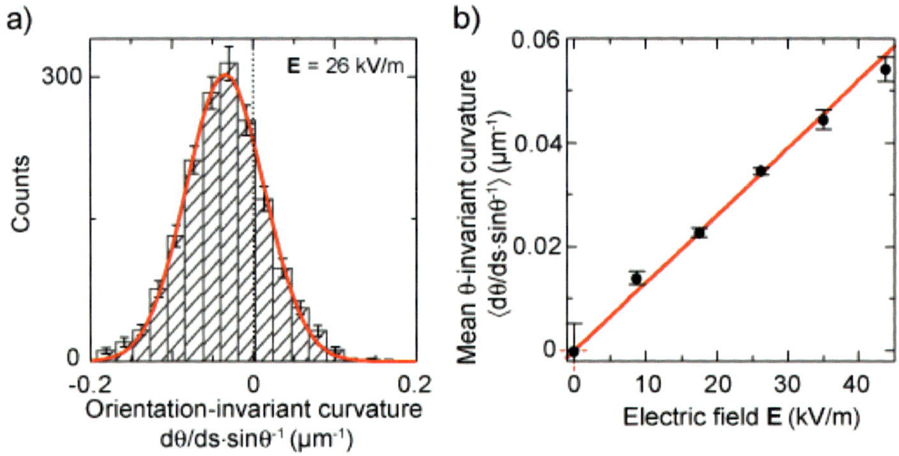

Figure 1.14 Quantification of the trajectory curvatures (a) Distribution of orientation-invariant trajectory curvatures measured from a large number of microtubules under an electric field of E = 26 kV/m. Red line is a Gaussian fit to the data. (b) Mean orientation-invariant curvature (taken as the centre of the distribution in panel a) varies linearly with the electric field, as expected from Equations 1.9 and 1.10. Adapted from reference [120] and reproduced with permission.

With this measurement of $\langle d\rangle$ we determine the persistence length of the leading microtubule ends as $p = 0.08 \pm 0.02$ mm. This value is much smaller than the persistence length of 4-8 mm that is measured for long microtubules.[125,126] Recent experiments have demonstrated that the persistence length of microtubules decreases from 5 mm to 0.11 mm upon decreasing the microtubule contour length from 48 μm down to 2.6 μm.[119] These observations were attributed to the anisotropic mechanical properties of microtubules, which are tubular structures consisting of 13 protofilaments arranged in parallel. In the proposed picture, sliding motion of neighbouring protofilaments induce a compliance in addition to the longitudinal stretching deformations of individual protofilaments. This leads to the expected decrease of the measured persistence length upon decreasing the deformation length.

An open question remained what happens for shorter lengths L of microtubules that could not be probed in the experiments in reference [119], because the observed $\propto L^{-2}$ decrease in persistence length predicts a vanishingly small persistence length on the $L \approx 0.1$ μm length scale that we probe in our experiments. In contrast, a recently proposed theoretical model, describing the bending of microtubules in terms of bundles of worm-like chains,[127] predicts a saturation of the persistence length upon a further decrease of the deformation length. Our method contributes a measurement of the persistence length of ~0.1 μm long tips, which is one order-of-magnitude smaller than the 2.6 μm length that was previously probed. Nevertheless, the value of $p = 0.08 \pm 0.02$ mm that we find is similar to the 0.11 mm that was measured for the 2.6 μm long microtubules. Moreover, in separate control experiments we established a tip persistence length of 0.24 ± 0.03 mm.[128] Thus our data indicate a lower bound on the persistence length of short lengths of microtubules which is consistent with a recently proposed theory describing the mechanics of wormlike bundles.

In conclusion, we have shown that electric forces in nanofabricated structures are an excellent tool for the study of the mechanical properties of individual biomolecules. We have measured the stiffness of short microtubule ends, which contributes to a better understanding of the mechanical properties of these macromolecules on short length scales.

1.4.3 Electrophoresis of Individual Microtubules in Microfluidic Channels

Finally, we describe the use of micron-size fluidic channels to confine and measure the electrophoresis of freely suspended individual microtubules.[122] Initially, these experiments were performed to measure the mobility of microtubules needed to calibrate the electric field-induced forces in steering experiments mentioned in the previous section. In addition, the high stiffness of microtubules makes them an excellent model system for rod-like particles, which provides an opportunity to measure and test the predicted anisotropy in the electrophoretic mobility for rod-like particles.[129] These experiments also allow us to measure the electrical properties of microtubules, such as the effective charge per tubulin dimer.

We observe the electrophoretic motion of fluorescently labelled microtubules inside 50 x 1 μm^2 slit-like channels that are fabricated between the entrance reservoirs that are separated by 5 mm. The experimental geometry is shown in Figure 1.11(c), with however one important difference: the omission of the kinesin molecules. Upon application of a voltage difference between the electrodes at either end of the channel, we observe that the freely suspended microtubules move in the direction opposite to the electric field. Note that the motion of the negatively charged microtubules in our channels is a superposition of their electrophoretic velocity and any fluid velocity inside the channel due to electro-osmotic flow.

In Figure 1.15(a) we show representative time-lapse images of two microtubules that are driven by an electric field $E = 4$ kV/m. The displacements of these microtubules, which are oriented with their axes approximately equal but in opposite directions to the field, are not collinear with the electric field. Instead, the velocity is slightly directed toward the axis of each microtubule. This orientation-dependent velocity is a hallmark of the anisotropic mobility of a cylindrical particle, Figure 1.15(b). The mobility μ of a microtubule is different for the electric field components perpendicular ($\mu_\perp$, $E_\perp$) and parallel ($\mu_{//}$, $E_{//}$) to its long axis. Consequently, a microtubule oriented under an angle θ with E (as defined in Figure 1.15(b)), will have velocity components parallel (v_y) and perpendicular (v_x) to the electric field:

$$v_x = \frac{1}{2}(\mu_{//} - \mu_{\perp})\sin(2\theta)E$$
$$v_y = \left[(\mu_{//} - \mu_{\perp})\sin^2(\theta) + (\mu_{\perp} + \mu_{EOF})\right]E \qquad (1.12)$$

where μ_{EOF} is the mobility of the electro-osmotic flow in our channels. We determine orientation-dependent velocity for a large number of microtubules. In Figure 1.15(c,d) we show binned values of measured v_x and v_y for microtubules at E = 4 kV/m. As expected from Equation 1.12, microtubules that are oriented under an angle with E move perpendicular to E in the positive x-direction if $\theta < 90^\circ$ and in the negative x-direction otherwise (Figure 1.15(c)). Moreover, microtubules that are oriented parallel to E ($\theta = 90^\circ$) move faster than microtubules that are oriented perpendicular to E ($\theta = 0^\circ$) (Figure 1.15(d)), which is expected if $\mu_{//} \geq \mu_{\perp}$ (Equation 1.12 and References 122 and 129). The red lines in Figure 1.15(c,d) are fits of Equation 1.12 to the data. The fitted amplitude A = $(\mu_{//} - \mu_{\perp})E$ and offset B = $(\mu_{\perp} + \mu_{EOF})E$ yield information about the different mobility components.

We measured orientation-dependent velocities for different electric fields and display the fitted A and B as a function of E in the insets of Figure 1.15(c,d). From the linear fit through the data we derive the values $(\mu_{//} - \mu_{\perp}) = -(4.42 \pm 0.12)\ 10^{-9}$ m^2/Vs, and $(\mu_{\perp} + \mu_{EOF}) = -(8.75 \pm 0.04)10^{-9}$ m^2/Vs. In order to determine the values of $\mu_{//}$ and $\mu_{\perp}$, we need to measure the value of the electro-osmotic flow mobility. We do this by a current-monitoring method[122,130] and we find $\mu_{EOF} = (1.28 \pm 0.01)10^{-8}$ m^2/Vs. This allows us to calculate $\mu_{//} = -(2.59 \pm 0.02)10^{-8}$ m^2/Vs and $\mu_{\perp} = -(2.15 \pm 0.01)10^{-8}$ m^2/Vs.

The measured mobility anisotropy $\mu_{\perp}/\mu_{//} = 0.83 \pm 0.01$ is clearly different from the well-known factor 0.5 in Stokes-drag coefficients for long cylinders. The reason is that in purely hydrodynamic motion, in which a particle is driven by an external force, the fluid disturbance around a particle is long-range, decaying inversely proportional to the characteristic length scale of the particle. However, in electrophoresis the external electric force acts on the charged particle itself, but as well on the counter ions around the particle. As a result the fluid disturbance around the particle is much shorter range and decays inversely to the cube of the characteristic length scale of the particle.[122,131]

The force on the counterions has important implications for the interpretation of electrophoresis experiments in terms of the effective charge. In previous reports of the electrophoretic mobility of microtubules, their motion was interpreted as a balance between the electric force on the particle and the hydrodynamic Stokes-drag coefficient.[117,132] However, because in hydrodynamic motion the fluid is sheared over a much larger distance than in electrophoresis, this interpretation seriously underestimates the restraining force, which leads to a similarly large underestimation of the effective charge.

Instead, we determine the effective charge of a microtubule by calculation of the ζ-potential and using the Grahame-equation that relates the ζ-potential to effective surface-charge density. For cylinders, the mobility $\mu_{//}$ is directly proportional to the ζ-potential via $\mu_{//} = \varepsilon\zeta\eta$,[129] where ε and η are the solution's dielectric constant and viscosity, respectively. This yields $\zeta = -32.6 \pm 0.3$ mV, which corresponds to an effective surface-charge density of -36.7 ± 0.4 mC/m^2. Using the surface area of a microtubule, we calculate an effective charge of -23 ± 0.2 *e* per dimer.[122] The latter value should be compared to the theoretical bare charge of -47 *e* per dimer, where we attribute the difference to screening due to immobile counter charges that are adsorbed to the microtubule between its surface and the no-slip plane. In contrast, previous reports that ignored the effect of counterions and thereby underestimated the restraining force[117,132] found a bare charge that was up to 5 orders of magnitude lower.

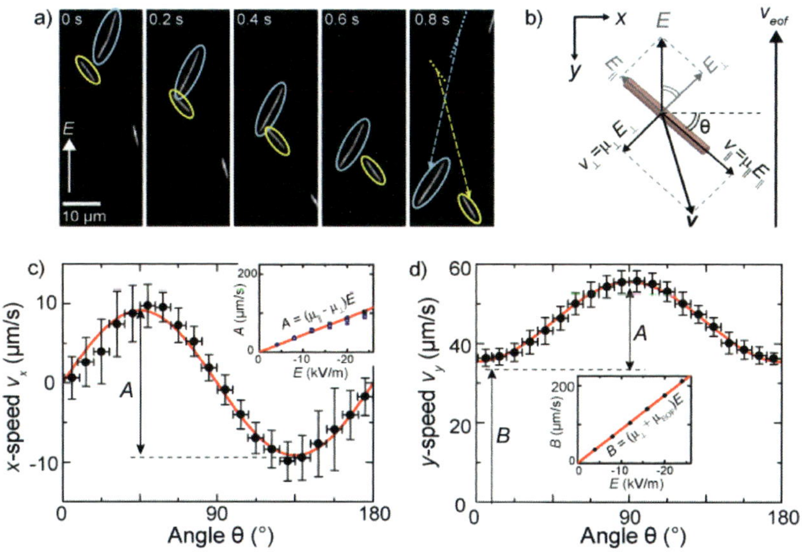

Figure 1.15 Electrophoresis of individual microtubules in microchannels. (a) Electrophoresis of microtubules under an angle with the electric field (E = 4 kV/m) is not collinear with the electric field. (b) A cylinder oriented under an angle θ with E and an anisotropic mobility for electrophoresis perpendicular ($\mu_\perp$) and parallel ($\mu_{//}$) to its axis will have a velocity v that is not collinear with E. The velocity will thus have components parallel (v_y) and perpendicular (v_x) to E (Eq. 1.12). (c) Measured v_x as a function of θ for microtubules at E = 4 kV/m. The solid line is a fit of Eq. 1.11. The inset shows the fitted amplitude A as a function of E. (d) Measured v_y as a function of θ for microtubules at E = 4 kV/m. The solid line is a fit of Equation 1.12. The inset shows the fitted amplitude offset B as a function of E. Adapted from reference [122] and reproduced with permission.

In summary, we have shown that microfabricated channels are an excellent system to measure the electrophoresis of individual microtubules. From these experiments we have gained valuable insights in the fundamental electrophoretic properties of colloidal cylinders and we obtained measurements of the effective surface charge of microtubules.

1.5 ACKNOWLEDGEMENTS

During the various projects we have benefited from the involvement and discussions with K. Besteman, D.J. Bonthuis, C.T. Butcher, Z. Deurvorst, R. Driessen, I. Dujovne, M.P. de Graaff, F.J.H. van der Heyden, W.J.A. Koopmans, M. Kruithof, S.G. Lemay, C. Meyer, Y. Shen, R.M.M. Smeets and with collaborators S. Diez, M. Dogterom, and J. Howard.

References

1. P. S. Dittrich, K. Tachikawa and A. Manz, *Anal Chem*, 2006, **78**, 3887-3907.
2. R. P. Feynman and J. Robbins, *The pleasure of finding things out: the best short works of Richard P. Feynman*, Perseus Books, Cambridge, Mass., 1999.
3. W. D. Volkmuth and R. H. Austin, *Nature*, 1992, **358**, 600-602.
4. S. W. Turner, A. M. Perez, A. Lopez and H. G. Craighead, *Journal of Vacuum Science & Technology B*, 1998, **16**, 3835-3840.
5. J. Han, S. W. Turner and H. G. Craighead, *Phys Rev Lett*, 1999, **83**, 1688-1691.
6. B. Hille, *Ion channels of excitable membranes*, 3rd edn., Sinauer, Sunderland, Mass., 2001.
7. T. E. Springer, T. A. Zawodzinski and S. Gottesfeld, *J Electrochem Soc*, 1991, **138**, 2334-2342.
8. J. Li, D. Stein, C. McMullan, D. Branton, M. J. Aziz and J. A. Golovchenko, *Nature*, 2001, **412**, 166-169.
9. C. Dekker, *Nature Nanotechnology*, 2007, **2**, 209-215.
10. H. A. Stone, A. D. Stroock and A. Ajdari, *Annual Review of Fluid Mechanics*, 2004, **36**, 381-411.
11. T. M. Squires and S. R. Quake, *Rev Mod Phys*, 2005, **77**, 977-1026.
12. H. Y. Wang, R. S. Foote, S. C. Jacobson, J. H. Schneibel and J. M. Ramsey, *Sensors and Actuators B-Chemical*, 1997, **45**, 199-207.
13. S. C. Jacobson, A. W. Moore and J. M. Ramsey, *Anal Chem*, 1995, **67**, 2059-2063.
14. F. H. J. van der Heyden, D. Stein and C. Dekker, *Phys Rev Lett*, 2005, **95**, 116104.
15. D. Stein, F. H. J. van der Heyden, W. J. A. Koopmans and C. Dekker, *Proceedings of the National Academy of Sciences of the United States of America*, 2006, **103**, 15853-15858.
16. F. H. J. van der Heyden, D. J. Bonthuis, D. Stein, C. Meyer and C. Dekker, *Nano Letters*, 2006, **6**, 2232-2237.
17. D. Stein, M. Kruithof and C. Dekker, *Phys Rev Lett*, 2004, **93**, -.
18. J. Lyklema, H. P. v. Leeuwen, M. v. Vliet and A. M. Cazabat, *Fundamentals of interface and colloid science*, Academic Press, London ; San Diego, 1991.
19. S. H. Behrens and M. Borkovec, *Phys Rev E*, 1999, **60**, 7040-7048.
20. A. Ajdari, *Phys Rev E*, 1996, **53**, 4996-5005.
21. S. e. a. Levine, *Faraday Transactions*, 1974, **2**, 1.
22. R. K. Iler, *The chemistry of silica : solubility, polymerization, colloid and surface properties, and biochemistry*, Wiley, New York, 1979.
23. H. Chang, F. Kosari, G. Andreadakis, M. A. Alam, G. Vasmatzis and R. Bashir, *Nano Letters*, 2004, **4**, 1551-1556.
24. R. Fan, M. Yue, R. Karnik, A. Majumdar and P. D. Yang, *Phys Rev Lett*, 2005, **95**, 086607.
25. R. M. M. Smeets, U. F. Keyser, D. Krapf, M. Y. Wu, N. H. Dekker and C. Dekker, *Nano Letters*, 2006, **6**, 89-95.
26. R. Karnik, K. Castelino, R. Fan, P. Yang and A. Majumdar, *Nano Letters*, 2005, **5**, 1638-1642.
27. R. B. Schoch and P. Renaud, *Applied Physics Letters*, 2005, **86**, 253111.
28. R. Karnik, R. Fan, M. Yue, D. Y. Li, P. D. Yang and A. Majumdar, *Nano Letters*, 2005, **5**, 943-948.
29. S. H. Behrens and D. G. Grier, *J Chem Phys*, 2001, **115**, 6716-6721.

30. J. N. Israelachvili and G. E. Adams, *Journal of the Chemical Society-Faraday Transactions I*, 1978, **74**, 975.
31. W. A. Ducker, T. J. Senden and R. M. Pashley, *Nature*, 1991, **353**, 239-241.
32. A. Naji and R. R. Netz, *Physical Review Letters*, 2005, **95**, 185703.
33. C. C. Fleck and R. R. Netz, *Physical Review Letters*, 2005, **95**, 128101
34. H. Boroudjerdi and R. R. Netz, *Journal of Physics-Condensed Matter*, 2005, **17**, S1137-S1151.
35. F. H. J. van der Heyden, D. Stein, K. Besteman, S. G. Lemay and C. Dekker, *Phys Rev Lett*, 2006, **96**, 224502.
36. J. Lyklema, *Journal of Physics-Condensed Matter*, 2001, **13**, 5027-5034.
37. L. Joly, C. Ybert, E. Trizac and L. Bocquet, *Phys Rev Lett*, 2004, **93**, 257805.
38. R. Qiao and N. R. Aluru, *Phys Rev Lett*, 2004, **92**, 198301.
39. A. Y. Grosberg, T. T. Nguyen and B. I. Shklovskii, *Reviews of Modern Physics*, 2002, **74**, 329-345.
40. Y. Levin, *Reports on Progress in Physics*, 2002, **65**, 1577-1632.
41. M. Quesada-Perez, E. Gonzalez-Tovar, A. Martin-Molina, M. Lozada-Cassou and R. Hidalgo-Alvarez, *Chemphyschem*, 2003, **4**, 235-248.
42. B. I. Shklovskii, *Phys Rev E*, 1999, **60**, 5802-5811.
43. K. Besteman, M. A. G. Zevenbergen, H. A. Heering and S. G. Lemay, *Phys Rev Lett*, 2004, **93**, 170802.
44. K. Besteman, M. A. G. Zevenbergen and S. G. Lemay, *Phys Rev E*, 2005, **72**, 061501.
45. P. Kekicheff, S. Marcelja, T. J. Senden and V. E. Shubin, *J Chem Phys*, 1993, **99**, 6098-6113.
46. M. Quesada-Perez, A. Martin-Molina, F. Galisteo-Gonzalez and R. Hidalgo-Alvarez, *Mol Phys*, 2002, **100**, 3029-3039.
47. T. Terao and T. Nakayama, *Phys Rev E*, 2001, **6304**, 041401.
48. Q. Wen and J. X. Tang, *J Chem Phys*, 2004, **121**, 12666-12670.
49. P. J. Scales, F. Grieser and T. W. Healy, *Langmuir*, 1990, **6**, 582-589.
50. J. F. Osterle, *ASME, TRANSACTIONS, SERIES E-JOURNAL OF APPLIED MECHANICS*, 1964, **31**, 161-164.
51. J. Yang, F. Z. Lu, L. W. Kostiuk and D. Y. Kwok, *Journal of Micromechanics and Microengineering*, 2003, **13**, 963-970.
52. H. Daiguji, P. D. Yang, A. J. Szeri and A. Majumdar, *Nano Letters*, 2004, **4**, 2315-2321.
53. W. Olthuis, B. Schippers, J. Eijkel and A. van den Berg, *Sensors and Actuators B-Chemical*, 2005, **111**, 385-389.
54. F. H. J. van der Heyden, D. J. Bonthuis, D. Stein, C. Meyer and C. Dekker, *Nano Letters*, 2007, **7**, 1022-1025.
55. E. Brunet and A. Ajdari, *Phys Rev E*, 2004, **69**, 016306.
56. J. Eijkel, *Lab on a Chip*, 2007, **7**, 299-301.
57. S. Pennathur, J. C. T. Eijkel and A. van den Berg, *Lab on a Chip*, 2007, **7**, 1234-1237.
58. Y. Ren and D. Stein, *Nanotechnology*, 2008, **19**, 195707.
59. E. Lauga, M. P. Brenner and H. A. Stone, in *Handbook of Experimental Fluid Dynamics*, eds. C. Tropea, A. Yarin and J. F. Foss, Springer, New York, Editon edn., 2005.
60. M. Majumder, N. Chopra, R. Andrews and B. J. Hinds, *Nature*, 2005, **438**, 44-44.
61. J. K. Holt, H. G. Park, Y. M. Wang, M. Stadermann, A. B. Artyukhin, C. P. Grigoropoulos, A. Noy and O. Bakajin, *Science*, 2006, **312**, 1034-1037.

62. B. Alberts, *Molecular biology of the cell*, 4th edn., Garland Science, New York, 2002.
63. G. W. Slater, P. Mayer and G. Drouin, *Analusis*, 1993, **21**, M25-M28.
64. J. L. Viovy, *Rev Mod Phys*, 2000, **72**, 813-872.
65. J. Han and H. G. Craighead, *Science*, 2000, **288**, 1026-1029.
66. M. Doi and S. F. Edwards, *The theory of polymer dynamics*, Oxford University Press, New York, 1986.
67. E. F. Casassa, *Journal of Polymer Science Part B-Polymer Letters*, 1967, **5**, 773.
68. E. F. Casassa and Y. Tagami, *Macromolecules*, 1969, **2**, 14.
69. E. A. DiMarzio and C. M. Guttman, *Macromolecules*, 1970, **3**, 131-146.
70. G. Taylor, *Proceedings of the Royal Society of London Series a-Mathematical and Physical Sciences*, 1953, **219**, 186-203.
71. H. Brenner, *Physicochem. Hydrodyn.*, 1980, **1**, 91-123.
72. P.-G. De Gennes, *Scaling concepts in polymer physics*, Cornell University Press, Ithica, 1979.
73. D. S. Cannell and F. Rondelez, *Macromolecules*, 1980, **13**, 1599.
74. Y. L. Chen, M. D. Graham, J. J. de Pablo, G. C. Randall, M. Gupta and P. S. Doyle, *Phys Rev E*, 2004, **70**, 060901.
75. J. O. Tegenfeldt, C. Prinz, H. Cao, S. Chou, W. W. Reisner, R. Riehn, Y. M. Wang, E. C. Cox, J. C. Sturm, P. Silberzan and R. H. Austin, *Proceedings of the National Academy of Sciences of the United States of America*, 2004, **101**, 10979-10983.
76. A. Balducci, P. Mao, J. Y. Han and P. S. Doyle, *Macromolecules*, 2006, **39**, 6273-6281.
77. D. Dutta and D. T. Leighton, *Anal Chem*, 2003, **75**, 57-70.
78. Y. C. Wang, A. L. Stevens and J. Y. Han, *Anal Chem*, 2005, **77**, 4293-4299.
79. J. P. Quirino and S. Terabe, *Science*, 1998, **282**, 465-468.
80. D. S. Burgi and R. L. Chien, *Anal Chem*, 1991, **63**, 2042-2047.
81. P. Gebauer, J. L. Beckers and P. Bocek, *Electrophoresis*, 2002, **23**, 1779-1785.
82. J. Astorga-Wells and H. Swerdlow, *Anal Chem*, 2003, **75**, 5207-5212.
83. S. Song, A. K. Singh and B. J. Kirby, *Anal Chem*, 2004, **76**, 4589-4592.
84. J. Khandurina, S. C. Jacobson, L. C. Waters, R. S. Foote and J. M. Ramsey, *Anal Chem*, 1999, **71**, 1815-1819.
85. D. Stein, Z. Deurvorst, F. J. H. v. d. Heyden, W. J. A. Koopmans and C. Dekker, *submitted*, 2008.
86. M. C. Williams, *Proceedings of the National Academy of Sciences of the United States of America*, 2007, **104**, 11125-11126.
87. S. Jun and B. Mulder, *Proceedings of the National Academy of Sciences of the United States of America*, 2006, **103**, 12388-12393.
88. D. J. Bonthuis, C. Meyer, D. Stein and C. Dekker, *submitted*, 2008.
89. J. H. Vanvliet, M. C. Luyten and G. Tenbrinke, *Macromolecules*, 1992, **25**, 3802-3806.
90. T. Odijk, *Polymer*, 1978, **19**, 989-990.
91. T. Odijk, *Macromolecules*, 1983, **16**, 1340-1344.
92. R. P. Feynman, *The pleasure of finding things out*, Penguin books, 1999.
93. J. Howard, *Mechanics of motor proteins and the cytoskeleton*, Sinauer Associates, Inc, Sunderland, Massachusetts, 2001.
94. J. M. Scholey, I. Brust-Mascher and A. Mogilner, *Nature*, 2003, **422**, 746-752.
95. H. C. Berg, *Annual Review Of Biochemistry*, 2003, **72**, 19-54.
96. R. D. Vale, *Cell*, 2003, **112**, 467-480.
97. R. D. Vale and R. A. Milligan, *Science*, 2000, **288**, 88-95.

98. K. Svoboda, C. F. Schmidt, B. J. Schnapp and S. M. Block, *Nature*, 1993, **365**, 721-727.
99. E. Meyhofer and J. Howard, *Proceedings Of The National Academy Of Sciences Of The United States Of America*, 1995, **92**, 574-578.
100. M. J. Schnitzer and S. M. Block, *Nature*, 1997, **388**, 386-390.
101. W. Hua, E. C. Young, M. L. Fleming and J. Gelles, *Nature*, 1997, **388**, 390-393.
102. D. L. Coy, M. Wagenbach and J. Howard, *Journal Of Biological Chemistry*, 1999, **274**, 3667-3671.
103. M. G. L. van den Heuvel and C. Dekker, *Science*, 2007, **317**, 333-336.
104. H. Hess, G. D. Bachand and V. Vogel, *Chemistry-a European Journal*, 2004, **10**, 2110-2116.
105. H. Hess, *Soft Matter*, 2006, **2**, 669-677.
106. R. Mukhopadhyay, *Analytical Chemistry*, 2005, **77**, 249A-252A.
107. S. Ramachandran, K. H. Ernst, G. D. Bachand, V. Vogel and H. Hess, *Small*, 2006, **2**, 330-334.
108. G. D. Bachand, S. B. Rivera, A. Carroll-Portillo, H. Hess and M. Bachand, *Small*, 2006, **2**, 381-385.
109. M. G. L. van den Heuvel, M. P. De Graaff and C. Dekker, *Science*, 2006, **312**, 910-914.
110. C. T. Lin, M. T. Kao, K. Kurabayashi and E. Meyhofer, *Small*, 2006, **2**, 281-287.
111. Y. Hiratsuka, T. Tada, K. Oiwa, T. Kanayama and T. Q. P. Uyeda, *Biophysical Journal*, 2001, **81**, 1555-1561.
112. S. G. Moorjani, L. Jia, T. N. Jackson and W. O. Hancock, *Nano Letters*, 2003, **3**, 633-637.
113. H. Hess, C. M. Matzke, R. K. Doot, J. Clemmens, G. D. Bachand, B. C. Bunker and V. Vogel, *Nano Letters*, 2003, **3**, 1651-1655.
114. M. G. L. van den Heuvel, C. T. Butcher, R. M. M. Smeets, S. Diez and C. Dekker, *Nano Letters*, 2005, **5**, 1117-1122.
115. M. Sundberg, J. P. Rosengren, R. Bunk, J. Lindahl, I. A. Nicholls, S. Tagerud, P. Omling, L. Montelius and A. Mansson, *Analytical Biochemistry*, 2003, **323**, 127-138.
116. D. Riveline, A. Ott, F. Julicher, D. A. Winkelmann, O. Cardoso, J. J. Lacapere, S. Magnusdottir, J. L. Viovy, L. Gorre-Talini and J. Prost, *European Biophysics Journal*, 1998, **27**, 403-408.
117. R. Stracke, K. J. Bohm, L. Wollweber, J. A. Tuszynski and E. Unger, *Biochemical and Biophysical Research Communications*, 2002, **293**, 602-609.
118. M. G. L. Van den Heuvel, C. T. Butcher, S. G. Lemay, S. Diez and C. Dekker, *Nano Letters*, 2005, **5**, 235-241.
119. F. Pampaloni, G. Lattanzi, A. Jonas, T. Surrey, E. Frey and E. L. Florin, *Proceedings Of The National Academy Of Sciences Of The United States Of America*, 2006, **103**, 10248-10253.
120. M. G. L. van den Heuvel, M. P. de Graaff and C. Dekker, *PNAS*, 2008, **in press**.
121. D. Stigter and C. Bustamante, *Biophysical Journal*, 1998, **75**, 1197-1210.
122. M. G. L. van den Heuvel, M. P. de Graaff, S. G. Lemay and C. Dekker, *PNAS*, 2007, **104**, 7770-7775.
123. A. J. Hunt and J. Howard, *Proceedings Of The National Academy Of Sciences Of The United States Of America*, 1993, **90**, 11653-11657.
124. T. Duke, T. E. Holy and S. Leibler, *Physical Review Letters*, 1995, **74**, 330-333.
125. M. E. Janson and M. Dogterom, *Biophysical Journal*, 2004, **87**, 2723-2736.
126. F. Gittes, B. Mickey, J. Nettleton and J. Howard, *Journal Of Cell Biology*, 1993, **120**, 923-934.

127. C. Heussinger, M. Bathe and E. Frey, *Physical Review Letters*, 2007, **99**.
128. M. G. L. van den Heuvel, S. Bolhuis and C. Dekker, *Nano Letters*, 2007, **7**, 3138-3144.
129. D. Stigter, *Journal Of Physical Chemistry*, 1978, **82**, 1417-1423.
130. X. H. Huang, M. J. Gordon and R. N. Zare, *Analytical Chemistry*, 1988, **60**, 1837-1838.
131. D. Long and A. Ajdari, *European Physical Journal E*, 2001, **4**, 29-32.
132. L. L. Jia, S. G. Moorjani, T. N. Jackson and W. O. Hancock, *Biomedical Microdevices*, 2004, **6**, 67-74.

CHAPTER 2

Biomolecule Separation, Concentration, and Detection using Nanofluidic Channels

JONGYOON HAN[1]

[1] Department of Electrical Engineering and Computer Science & Department of Biological Engineering, Massachusetts Institute of Technology, 77 Massachusetts Ave. Cambridge, MA 02139, USA

2.1 INTRODUCTION

In recent years, science and engineering involving very small scale fluidic channels (nanofluidic channels) has attracted much attention. Many nanofluidic systems have been developed and reported with a continual increase in the number of papers that incorporate 'nanofluidic phenomena'. Many have asked the question, *Is this the beginning of a new field of research, or just a natural evolution of the better-established field of microfluidics, or just another short-lived fad?* Indeed, why do we want to study nanofluidics, and why should we develop nanofluidic systems? What is the intellectual core of this new area of research? Is this a new science or is this engineering?

Many excellent review papers have already published on this topic, and essentially time will answer all of the above questions, as the field naturally evolves. However, at the risk of being redundant, I would like to discuss some reasons why I believe that the field of nanofluidics was and still is, stimulating, relevant and (potentially) useful.

1) *Because we have already been using it for a long time*

Nanofluidic systems have been used for a long time in a diversity of industries and applications and in manipulating small particles, molecules and fluids. At the time, we simply did not call them "nanofluidic systems" or give them any particular name for that matter. The common features of all these systems are the nanostructures made to affect the transport of molecules and nanoscale objects. Various nanoporous membranes, gels and molecular filters have been widely used in chemical engineering, biology and chemistry. For example, renal filtration in the kidney is the ultimate filtration system for our body; the failure of which produces dire consequences. Additionally, biological membranes in cells employ very efficient nanopores in order to transport various molecules.

2) *Because we don't know how it really works*

In spite of this ubiquity of 'nanofluidic' systems around us, our understanding of these systems is actually quite limited. The main reason for such a limited understanding is that traditional nanoporous/membrane systems have essentially random structures. Without the key structural parameters such as pore size, engineers and scientists cannot really build an engineering model that is based on first principles. Instead, attempts to model nanoporous gels and membranes have been largely phenomenological, incorporating many adjustable parameters. Nanofluidic channels provide ideal experimental platforms to test all these models for membranes and nanoporous systems and promote further understanding of membrane phenomena.

3) *Because we can make nanofluidic components*

Recent advances in micro/nanofabrication methods have allowed one to fabricate nanofilters and nanopores with critical dimensions of a few nanometers. Therefore, the engineering and scientific communities have taken advantage of such an ability to discover, identify, and demonstrate novel and rediscovered phenomena at such length scales. Significantly, some of these phenomena have turned out to have practical implications in various applications. Today it is possible to routinely fabricate structures as small as molecules themselves, in the size range of 1~10 nm.

4) *Because it might be useful for various applications*

There is a great interest in mimicking and even replacing naturally occurring systems such as the kidney. Even with a sustained effort, we still cannot fully replace the (amazing) functionality of a kidney, which is a big problem for biomedical engineers. Additionally, some of the phenomena discovered in nanofluidic systems could be utilized for new types of filters and membranes.

In my opinion, nanofluidics is certainly not an entirely new idea or research field, but rather a re-discovery of existing problems, in the context of the recent developments in nanofabrication and MEMS technology. The scientific issues involved in this area, such as hydrodynamics at the nanoscale, molecular confinement, and complex sieving phenomena, are ones that have been studied for many years, in various fields of science and engineering. One potential benefit of this new area of research will be the contribution to scientific understanding of these problems, aided by experimental tools provided by nanofluidic filters, channels and pores. The various biomedical applications suggested and demonstrated in this area will likely be the main drivers for development.

The predominant aim of this chapter is to provide a summary of the important engineering aspects relevant to the field of nanofluidics, with a particular focus on biomolecule separation, molecular concentration and molecule detection. For a discussion of additional aspects of nanofluidics, the reader is directed to the many excellent reviews in this book.

2.2 FABRICATION TECHNIQUES FOR NANOFLUIDIC CHANNELS

2.2.1 Etching & Substrate Bonding Methods

One of the simplest methods to realize nanometer scale fluidic channels involves etching very shallow fluidic channels using standard reactive ion etching (RIE), and closure by bonding to another flat substrate. A nanochannel made in this method would be nanometers deep, but much wider in the lateral dimension. Since many nanofluidic applications require molecular/fluidic confinement only in one dimension, vertical confinement provided by such a fabrication technique is useful for many different molecular filtration and separation applications. The advantage of this approach is that highly regular nanofluidic channels with good dimensional control can be realised utilizing only standard photolithography and bonding methods. In other words, there is no need for advanced nanolithography tools to create nanostructures. The only two important considerations relating to this technique are:

1) One needs to fine tune the bonding process carefully, so as not to collapse the nanochannels created by etching. This is important because most substrate bonding processes (thermal or field-driven bonding processes) occur at temperatures near the glass transition temperature of the substrate. While higher bonding temperatures typically lead to superior bonding strength, excessive bonding temperatures will induce the collapse of the nanochannel ceiling. Mao *et al.* characterized detailed bonding process parameters for creating regular nanochannels, using both anodic bonding (glass-silicon, field driven bonding) and thermal bonding (glass-glass wafers).[1] In both cases, bonding temperatures and the aspect ratio (channel width/channel depth) can be controlled to generate nanochannels with a depth of 20nm or less (Figure 2.1(A)). In addition, the fabrication of nanochannels in plastic substrates using a similar etch-bond method has recently been reported.[2] Moreover, novel processes for structuring even thinner nanochannels using similar techniques have also been reported.[3] These developments have led to new scientific discoveries on fluidic filling in nanofluidic channels and other phenomena, which are not usually found in microfluidic channels.

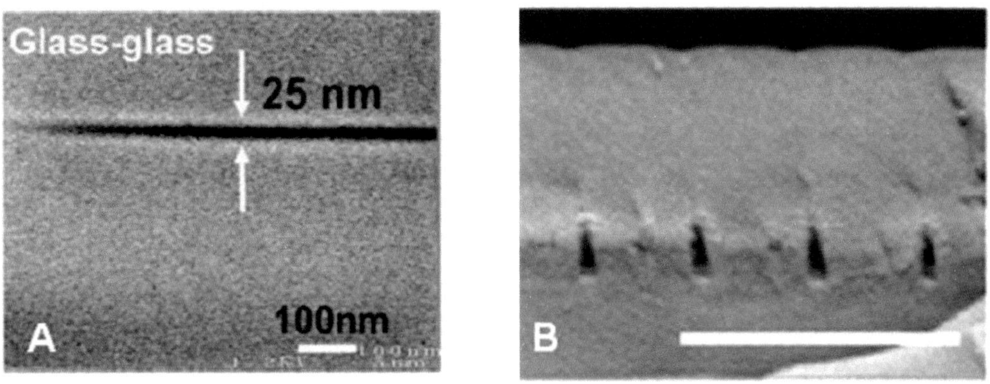

Figure 2.1 (A) 25nm thin, two-dimensional planar nanochannel, adapted from reference [1] and adapted with permission. (B) An array of one-dimensional, linear nanochannels with the critical dimension of ~17nm (scale bar: 500nm). Adapted from reference [8] and reproduced with permission.

2) The technique tends to generate 'planar' nanochannels, with a relatively low sample and fluid throughput. Whilst this is not a serious problem when performing many scientific studies, it becomes a particularly important issue when designing practical nanofluidic molecular sieving/filtration devices. In order to address this issue, Mao *et al.* have reported a fabrication technique for making massively parallel nanofilters utilizing the anisotropic etching of silicon.[4]

2.2.2 Sacrificial Layer Etching Techniques

Another method for generating nanochannels involves the use of sacrificial layers, Such an approach has been widely-used for creating suspended MEMS structures. Fluidic channels (nanochannels and microchannels) can be defined by depositing a sacrificial layer which is sealed with another capping layer. Subsequently, the sacrificial layer is removed by a highly-selective etching process. The greatest advantage of this technique is that there is no need for substrate-bonding (which is often very sensitive to wafer flatness), since the sealing of the fluidic channel is achieved as part of processing. This approach can allow fabrication of complex, three-dimensional fluidic networks with multiple channels crossing each other. Disadvantages of the methodology include extended etching times (especially when the enclosed fluidic channels are long) and defect generation due to stresses on the capping layer.

Since the thickness of the deposited sacrificial layer can be controlled very accurately, one can make highly thin, planar nanochannels using this method.[5,6] For example, Turner *et al.* used poly-Si as a sacrificial layer to build complex nano/microchannels, with elaborate openings and supporting structures that prevent channel collapse due to stress.[7] Additionally, Cao *et al.* used nanoimprint lithography and non-uniform deposition techniques to generate sealed nanochannels as small as 10 nm×50 nm.[8] (Figure 2.1(B))

2.2.3 Other Fabrication Methods

The issue of limited sample throughput can only be resolved by creating parallel arrays of nanopores / nanofilters, either by relying on polymeric nanoporous structures (such as gels) or relying on alternative, non-standard fabrication methods. Many different techniques to build regular nanopore / nanochannel structures with good pore size control have been developed. Nuclear track-etch nanopore membranes,[9-11] and the honeycomb structure of Anodized Alumina Oxide (AAO)[12] are the two well-known techniques. Non-traditional methods have also been used to generate the same effect. For example, Jeon *et al.* combined a PDMS phase mask and photopolymerization to create three-dimensional submicron pore systems.[13] Alternatively, close-packed micro- or nanoparticles could be used as a regular nanoporous structure, or as a template to fabricate one.[14] These techniques are generally economical methods for generating parallel nanoporous systems, but their integration into monolithic systems is challenging, although recent efforts have been quite successful.[15,16] In addition, only periodic structures can be created using these techniques, with somewhat limited long-range order.

2.3 BIOMOLECULE SEPARATION USING NANOCHANNELS

2.3.1 Molecular Sieving using Nanofluidic Filters

One obvious application of nanochannels is their use as a filter for molecules and particles. In fact, this was the first application (and motivation) for building nanofluidic structures, i.e. 'artificial gels' for biomolecule separation.[17] While it is now possible to make a nanofilter or nanopore as large as a single protein, the 'filtration' phenomena that occur within a nanofilter can be quite complicated. Biomolecules are usually flexible 3D structures with internal degrees of freedom (conformation). Moreover, stochastic motion of biomolecules (Brownian motion) has a significant impact on the way that molecules interact with the nanopores within a nanofilter. While there are more detailed reviews on the different sieving regimes,[18] examples of nanofluidic molecular sieving structures for different sieving modes are briefly reviewed.

Polymeric biomolecules such as DNA and RNA form a spherical 'blob' conformation in a liquid because the free energy for an extended conformation is much higher. Even though the width of such biopolymers is small (2-3 nm), the relevant size for molecular sieving is the radius of gyration (R_g), which is the approximate radius of the spherical blob conformation. For example, a λ-DNA molecule (48.5 kbp) has an R_g of ~0.7 μm. When such a molecule is driven into a nanofilter with a filter gap size much smaller than R_g, the molecule will get stuck (or entropically trapped) when the driving electric field is not sufficient to overcome the entropic force (which keeps the spherical conformation of the polymer).[19] DNA molecules escape these 'entropic traps' only when the driving electric field is sufficient to stretch the DNA. Even then, longer molecules have a larger probability of escaping the entropic traps, contrary to usual intuition. This is due to the fact that the stretching of DNA (which is critical in the escape process) is initiated by local deformation, which is independent of the overall blob size.[19] Using this approach, Han and Craighead separated large DNA molecules with high efficiency.[20]

For globular biomolecules, such as proteins or shorter DNA and RNA molecules, conformation becomes less important in sieving, and such molecules can be approximated as semi-rigid particles. Indeed several decades ago, it was both experimentally[21] and theoretically[22] established that molecules of this kind will be sieved and filtered from entering a nanopore or nanochannel, even when the channel opening is larger than the size of the molecule. Therefore, it is possible to separate proteins and other small biomolecules (a few nm in size) with relatively large nanofilters, as demonstrated by Fu and co-workers.[23] While this is exactly the same mechanism by which proteins and small DNA molecules are separated in polyacrylamide gels, experiments in gels cannot be readily used to test Ogston sieving models due to the uncertainty and randomness of gel pore sizes. Nanofilters, on the other hand, can be made to have a well-defined pore size and shape. Fu *et al.* developed a simple theoretical model based on Kramer's theory, which enabled a detailed characterization of field-dependent electrophoretic mobility in Ogston sieving by nanofilter arrays.[24] Such an 'anomalous' field-dependent mobility has long been observed in gel electrophoresis experiments,[25] with the traditional near-equilibrium Ogston sieving model unable to quantitatively describe such behaviour.

In practical engineering applications, more emphasis is put on separation throughput and speed rather than separation resolution, especially in non-genomic applications. This is mainly due to the fact that most biological samples (e.g. blood and saliva) are highly complex in terms of biomolecule composition, containing entities of widely varying size (from small molecules to cells) and concentration (mM - pM). On the other hand, detection systems such as immunoassays and mass spectrometry allow one to discern target biomolecules over a molecular background. Traditional elution-type microfluidic separation systems (essentially a miniaturization of capillary electrophoresis) are not ideal in this regard, since one can only process very small volumes (pL - nL) at any time. As an alternative, various types of continuous-flow separation systems are actively being developed. Especially for nanofilter-based separation systems, a continuous-flow separation mode is beneficial because one can significantly reduce the chance of clogging by larger particles.

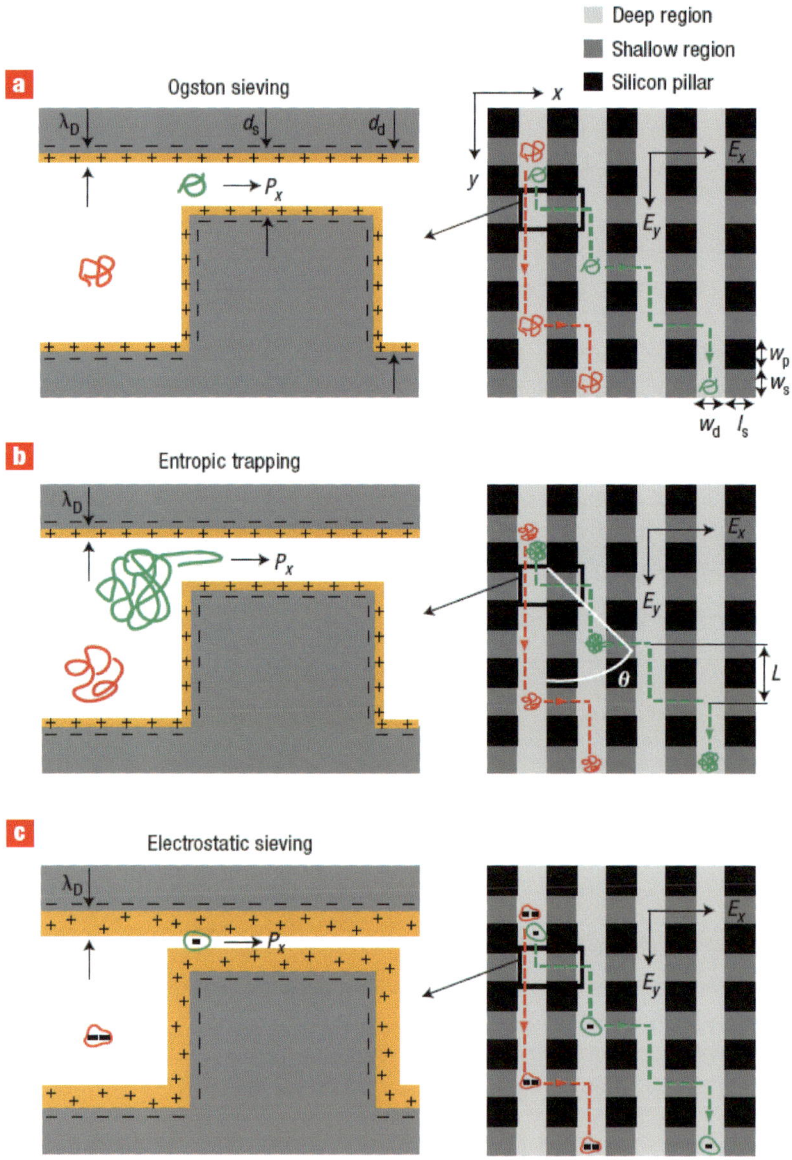

Figure 2.2 Three different sieving modes in a nanofluidic filter array. (a) Ogston sieving applies for molecules (both polymeric and globular) smaller than the nanofilter gap size. (b) Entropic trapping applies for long polyelectrolytes (much longer than the persistence length of the polymer backbone), when the radius of gyration larger is than the nanochannel gap size. (c) Electrostatic sieving applies when the Debye length within the nanochannel is comparable to the nanochannel gap size, rendering transport through the nanochannel charge-selective. Adapted from reference [42] and reproduced with permission.

For example, Fu and co-workers demonstrated a continuous-flow nanofilter array system, where proteins and DNA molecules may be separated using many different sieving mechanisms including Ogston sieving, entropic trapping, and electrostatic sieving.[26] (Figure 2.2) Electrostatic sieving occurs when the Debye layer essentially spans the entire depth of a nanochannel, by decreasing the ionic strength. Then molecular transport through the nanochannel will be perm-selective, meaning that a protein's mobility through the nanochannel will depend on the charge density (or pI value of the protein). A similar idea can also be used to decrease the effective size of the nanofilter (therefore leading to higher selectivity), since counter-ions (negatively charged proteins) will be prohibited from entering the non-electroneutral Debye layer.

2.3.2 Computational Modelling of Nanofilter Sieving Phenomena

When using regular nanofilters, one can study the detailed physical phenomena of molecular sieving in a controlled experiment. The results of molecular sieving studies involving nanofilter-generated data often surprise researchers, signifying that the physical problem at hand is not as simple as the intuitive filtration model. This is especially true for long polymeric biomolecules, where conformational degrees of freedom complicate the molecular interaction with the nanofilter.[19,27] Due to this complexity, there have been many attempts to simulate or model polymer interactions with nanofilter sieving structures. Techniques used range from simple analytical models such as Transition State Theory,[19, 24] Monte Carlo simulations[28] and Brownian Dynamics simulations.[29,30] While these models reproduce and simulate molecular sieving events in nanofilters, we have yet to see the development of numerical models that can be used for predicting and optimizing nanofilter sieving systems.

The main difficulty in computational modelling is that accurate description of the problem requires the modelling of both stochastic dynamics of biomolecules and molecular hydrodynamics. While most simulations (Brownian Dynamics and Molecular Dynamics) focus on stochastic motion of biomolecules near the nanofilter, it is well known that charged molecules trapped in the nanofilter will generate local electroosmotic flow, due to the presence of counterions.[25] Such local molecular hydrodynamics is also important in determining the diffusivity of the molecules (Zimm diffusivity), as well as band dispersion of the system.

Two recent approaches are noteworthy in this respect. Li *et al.* recently developed a continuum transport model for Ogston sieving, where the detailed Brownian dynamics of rigid biomolecules near a nanofilter were averaged into an effective entropic potential term, in order to simplify the model by eliminating the need for stochastic simulation.[31] In this situation, one can simply solve a standard continuum transport equation, with full consideration of any hydrodynamic flow in the system. As a result, both separation selectivity and peak dispersion can be modelled for system optimization. Moreover, Duong-Hong *et al.* applied Dissipative Particle Dynamics to model both the stochastic motion of DNA and its hydrodynamics in entropic trapping system.[32,33] Such coarse-grained stochastic simulation tools include all the relevant physics of the problem, whilst allowing one to analyze even mesoscale nanofilter devices, which are too expensive to model in standard molecular dynamics techniques.

2.4 BIOMOLECULE CONCENTRATION USING NANOCHANNELS

2.4.1 Biomolecule Pre-concentration using Nanochannels and Nanomaterials

One of the fundamental problems in molecular medical diagnostics is that biomarkers (whose concentration may correlate with disease progress) are generally low-abundance proteins in bio-fluids such as blood or saliva, and often below the detection limit of conventional biosensors. The lack of effective molecular amplification strategies (such as PCR) for non-nucleotides has been the major bottleneck in protein biosensing. One viable strategy to address this issue is the pre-concentration of dilute samples into smaller sample volumes having higher concentrations. Conventional methods for pre-concentration involve affinity-based biomolecule traps, which provide detection sensitivity enhancements of between one and two orders of magnitude. The main disadvantages of affinity-based pre-concentration include the requirement of buffer release (to dilute the concentrated plug), sample loss, and limited pre-concentration capacity (determined by the binding surface).

Nanochannels or nanoporous materials have been demonstrated as an effective protein pre-concentration tool. Various mechanisms can be utilized for this purpose. For example, Wang *et al.* used a perm-selective nanochannel (~40nm) to induce strong ion concentration polarization in the vicinity of nanochannel junction.[34] (Figure 2.3) On the anodic side of the junction, ions and other charged biomolecules are repelled (depleted) from the nanochannel, which is often referred to as ion depletion. This force was matched with additional flow in the opposite direction to create a continuous biomolecule trap. This pre-concentration device can be turned on and off using the field and can be operated continuously until enough pre-concentration is achieved. Recently, similar devices were implemented in a PDMS microfluidic platform, utilizing various methods of creating nanofluidic junctions.[35-38] (Figure 2.4, top)

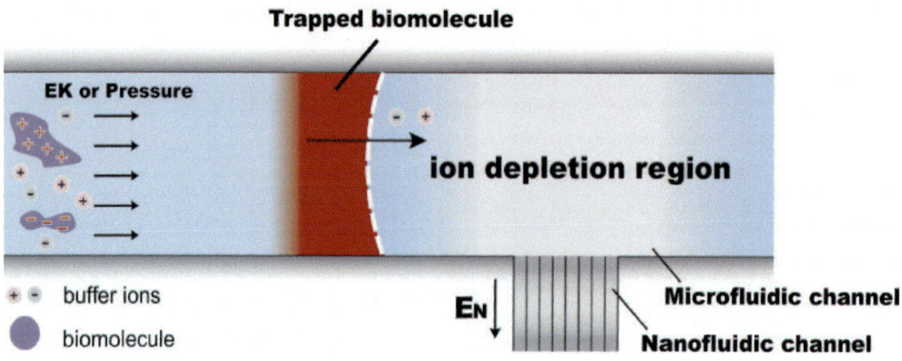

Figure 2.3 Schematic and mechanism of a nanofluidic biomolecule pre-concentration device. Adapted from Novel Nanofluidic Electrokinetic Trapping Device for Proteomic Applications, Y-C. Ph.D. Thesis, Massachusetts Institute of Technology, 2007

In the ion concentration polarization near perm-selective nanochannels, the cathodic side of the junction experiences ion enrichment,[39] which can also be used for biomolecule pre-concentration.[40] In addition, nanofilters and nanoporous materials can be used as a filtration membrane, which will pre-concentrate molecules larger than the pore size. Both silica membranes[41] and photo-patterned gels[42,43] have been used for this purpose (Figure 2.4).

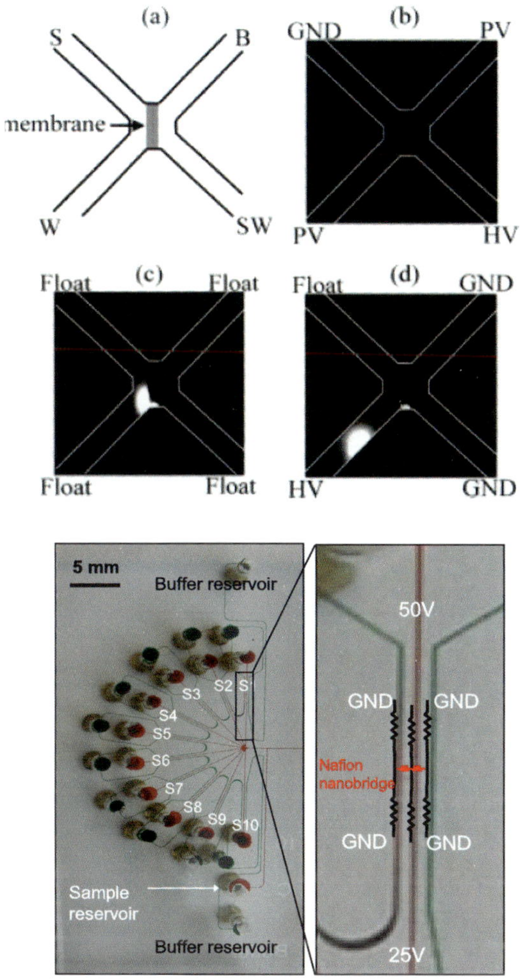

Figure 2.4 Various pre-concentration systems using nanoporous materials. *Top*: Photopatterned polyacrylamide membrane used in a filtration pre-concentraion of proteins. Adapted from reference [42] and reproduced with permission. *Bottom*: Array of ten electrokinetic pre-concentrators fabricated by surface patterning of Nafion® resin. Adapted from reference [37] and reproduced with permission.

These microfluidic pre-concentration systems are especially useful for enhancing the sensitivity and selectivity of microfluidic immunoassays. The sensitivity of standard immunoassays is largely determined by the quality of the antibodies used, and more specifically the dissociation constant (K_D) of the antigen-antibody couple. When the target molecule's concentration is much lower than K_D (which is often the case for biomarkers in blood serum) primary immuno-binding of targets to the antibody is slow and inefficient, leading to a noisy sensor signal. Nanochannel pre-concentration systems have been successfully integrated with immunoassays in microfluidic channels to enhance both the detection sensitivity and dynamic range (Figure 2.5).[44,45] One important aspect of this approach is that the sensitivity enhancement is independent of the specific type of immunosensor used, since the enhancement is obtained by sample concentration. Theoretically, any conventional or novel immunosensor could be integrated with a pre-concentration system for improved performance. In addition, other kinds of biochemical reactions (not only the binding reaction between an antibody and antigen) can also be enhanced by concentrating the two reactants. For example, Lee *et al.* used a pre-concentration device to enhance the standard enzyme activity assay, by concentrating both target enzymes (the activity of which to be measured) and fluorogenic substrates.[46] Using this, the detection sensitivity of the activity assay was improved by almost three orders of magnitude. Such advancements may lead to more sensitive detection of low-level enzymes in bio fluids in the future.

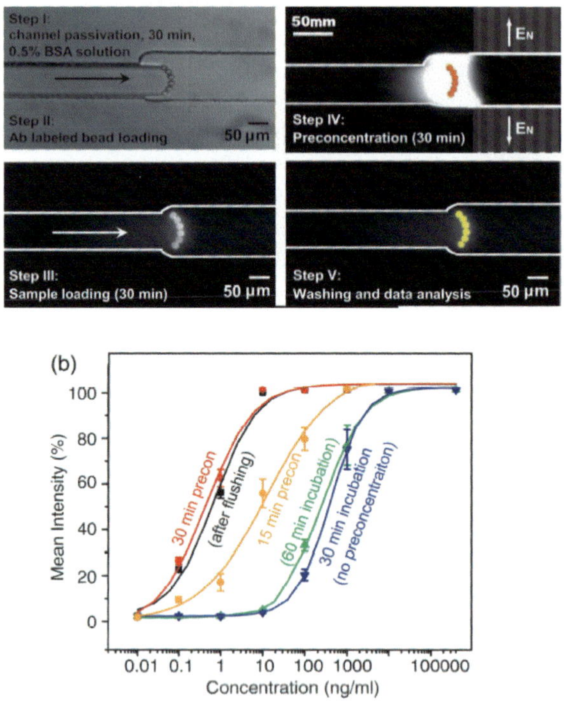

Figure 2.5 (Top) Integration of nanofluidic pre-concentrator with bead-based immunoassays. (Bottom) Plots showing the dose response of an immunoassay without pre-concentration, and with 15 min or 30 min pre-concentration. Through maintaining a 30 min on-site pre-concentration, this approach can lower the sensitivity limit by about 500 fold from 50 pM to sub 100 fM range (with 10 μg ml^{-1} GFP as simulated molecular background). Adapted from reference [45] and reproduced with permission.

2.4.2 Non-Linear Electrokinetic Phenomena near Nanochannels

The ion concentration polarization phenomenon, which occurs near relatively thin nanochannels at low to moderate ionic strengths (even up to ~10mM), is indeed quite a general phenomenon. It has been shown that even ion currents through polyacrylamide gel[43] or weakly bonded PDMS-glass gaps[35,36] could induce ion concentration polarization. This phenomenon has been studied in membrane science, where a perm-selective membrane such as Nafion® is used in applications such as fuel cells and desalinization. Classical theory of ion concentration polarization predicts that, for cation-selective membranes (negatively charged, therefore preferentially transfer cations), ions will accumulate on the cathodic side (ion enrichment) and will be depleted on the anodic side (ion depletion).[47] The driving force for this process is essentially the electrostatic interaction between charges, driving the system towards electroneutrality. When ions are depleted and enriched in the system, the electrokinetics become extremely complicated since a uniform zeta potential (and Debye length) across the system cannot be assumed, and thus the full Poisson-Nernst-Planck equations must be solved. Qualitatively speaking, ion depletion makes the anodic side of the nanochannel exhibit a lower ionic strength, leading to a stronger electroosmotic response at a given applied field. Rubinstein and co-workers have theoretically suggested that there exists a strong convective mixing (electroosmotic flow of the second kind, which destroys the concentration polarization), created by the 'amplified' electrokinetic response of the fluid layer right next to the membrane.[48] Kim, Wang and co-workers experimentally visualized the nonlinear electrokinetic flow inside and outside the ion depletion region by tracking various charged fluorescent dyes and particles *in situ*.[49] They also measured the current-voltage behaviour through the nanochannel junction, along with the concentration development of the system (Figure 2.6). The current profile can be divided into three regimes similar to ones reported in perm-selective membrane studies. Following the Ohmic regime, the onset of the concentration polarization phenomenon concurs when limiting currents are measured. Moreover, if one further increases the applied potential, the depletion region can be additionally expanded and the current extended beyond the limiting condition. When a strong, vortex-like flow is observed, the ion-depletion and pre-concentration process is not efficient, presumably due to the strong convective mixing destroying the ion depletion region. Such a simultaneous measurement of current, concentration and flow near the micro-nanometer junction, with optical (micron) resolution has been enabled by the nanofluidic channel systems and is a significant improvement in our understanding of various membrane phenomena.

2.5 CONFINEMENT OF BIOMOLECULES USING NANOCHANNELS

2.5.1 Nanochannel Confinement of Biomolecules

In experimental biochemistry, activities and reactions of biomolecules are characterized mostly in a test tube, and the resulting free solution reaction properties are used in understanding biochemical reactions within cells and tissues. However, most interesting biochemical reactions do not occur in a free solution, but in a confined space or on a surface. The cell cytoplasm is far from a free solution and closer to a polymer melt solution, crowded with many proteins and organelles. Many critical biochemical processes occur in the endoplasmic reticulum membrane, which consists of slit-like spaces. Extracellular cell signalling molecules should be transported through the nanoporous extracellular matrix (e.g. glycocalyx), eventually becoming bound to cell surface receptors. Whether reactions in such confined environments can be correctly described by free-solution measurements is therefore a valid and important question.

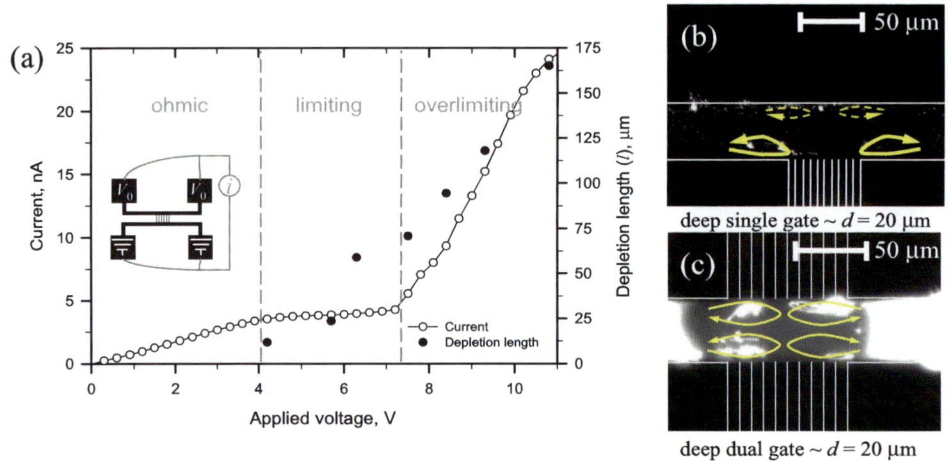

Figure 2.6 Non-linear electrokinetic response in a nanofluidic channel. (a) Current sweep plot showing the Ohmic-current, the limiting-current and over-liminting current pattern. (b) Fast vortex at steady state in a single gate (SG) device and (c) four independent strong vortices in dual gate (DG) device. Adapted from reference [49] and reproduced with permission.

If polymeric biomolecules such as DNA or RNA are confined in a tight space smaller than the radius of gyration of the molecule, their physico-chemical properties will be significantly altered. This problem has been studied in theoretical polymer physics since the 1970s, and nanochannels provide an ideal experimental platform to study such problems. For example, using planar, slit-like nanochannels (having a gap size much smaller than the radius of gyration of DNA), it was shown that DNA polymer relaxation could be significantly slowed down in a confined space.[50] Within a planar nanochannel, a DNA blob (a near spherical conformation of DNA) will be squeezed, and the diffusivity and electrophoretic mobility of the DNA will be significantly modified from the bulk value. Balducci *et al.* measured the diffusivity of long DNA molecules in planar nanochannels and compared their results with previously known scaling laws.[51]

It is also possible to confine DNA molecules within thin and narrow nanochannels (one-dimensional channels). Since the persistence length (the length scale below which DNA molecules can be considered as a rigid rod) of double stranded DNA is about 50 nm, confinement of a long DNA strand in a one-dimensional nanochannel comparable to this length scale will essentially make the DNA completely stretched (Odjik limit). Tegenfeldt *et al.* measured the conformation and dynamics of long DNA molecules within one-dimensional nanochannels and compared the experimental results with theory.[52,53] Using similar one-dimensional nanochannels, Wang *et al.* studied the biochemical reaction between DNA and DNA-binding proteins.[54] These studies are important and relevant in understanding DNA-protein interactions *in vivo* and provide a solid biophysical foundation for characterizing biochemical reactions in confined spaces.

The controlled stretching of long DNA strands has another important application in genome mapping.[55] However, one bottleneck in this situation is the ability to overcome the entropic force (which keeps the DNA polymer in a spherical shape), especially at the nanochannel/microchannel junction. In fact, the final stretched conformation of a DNA strand can vary markedly due to so-called molecular individualism.[56] Several different strategies have been suggested. For example, Cao *et al.* designed a structure containing a gradient transition from large (micron sized) to small (nanometre sized) pores.[57] Moreover, Randall *et al.* designed a hyperbolically converging microchannel to continuously stretch long DNA molecules for downstream DNA mapping.[58]

2.5.2 Enhancement of Binding Assays using Molecule Confinement in Nanochannels

Nanochannel confinement of biomolecules has another interesting application in biosensing and biochemical processing. One limiting factor in low-abundance analyte detection by immunoassays (such as ELISA) is the existence of surface diffusion layers,[59] which limit binding kinetics. Convection can be used to overcome this transport limitation, but only up to a certain level.[60] This is mainly due to the fact that the diffusion timescale ($\sim L^2/D$) becomes much shorter than the convection timescale ($\sim L/v$, where v is the velocity of the flow), especially when L becomes small. Therefore, it is very difficult to 'disrupt' a thin diffusion layer, even with relatively strong flow.

Nanochannels, however, can force target molecules to be very close to the binding partner on the surface, by confining the target molecules. Simply put, within a nanochannel the diffusion length is limited by the depth of the nanochannel, which can be 50 nm or less. Diffusive transport over such a short distance is very efficient, and therefore surface binding reactions can occur very efficiently in nanochannels. This effect was utilized by Schoch *et al.* to demonstrate enhanced binding reaction rates within nanochannels.[61] The internal surface of a relatively short nanochannel was coated with specific antibody to the target. A sample solution containing the target molecule at low concentration was driven through the nanochannel by pressure driven flow. In nanochannels, diffusive transport still dominates even at the relatively high flow speed used (2-3 cm/s). Consequently, almost all the target molecules that enter the nanochannel will get bound to the antibody, and the binding reaction speed will be determined by the flow speed. By inducing fast flows, Schoch *et al.* demonstrated that the reaction rate could be increased by a factor of 50, compared with the previously reported diffusion-limited reaction in nanochannels.[62] This work provides an interesting and important insight not only in biosensing and chemical processing, but also in highlighting the difference between free-solution binding kinetics and those in confined spaces such as membranes.

2.6 CONCLUSIONS AND FUTURE DIRECTIONS

In this chapter, various recent advances in nanofluidic engineering and science have been reviewed. A range of microfluidic structures have been actively designed and fabricated in order to manipulate and control cellular systems in a manner that has not been possible with conventional cell culture techniques. It is highly likely that various nanofluidic structures, with critical dimensions ranging between 1 and 100nm, made by both conventional and non-conventional fabrication techniques, will play a similar role in promoting our understanding of various biomolecules. Such structures (nanoporous materials and filters) have indeed been used for a long time in various applications, but with limited understanding of the precise physical phenomena in or near the nanopore. Nanochannels are likely to contribute to scientific advances in those fields, by providing informative model membranes and filters with the capacity to track the motion of ions, molecules and fluid *in situ*.

2.7 ACKNOWLEDGEMENTS

In preparing this review, the author acknowledges support from NIH (EB005743), KIST-IMC (Korea), and Singapore-MIT Alliance (SMA-II).

References

1. P. Mao and J. Han, *Lab Chip*, 2005, **5** 837 - 844.
2. P. Abgrall, L.-N. Lowb and N.-T. Nguyen, *Lab Chip*, 2007, **7**, 520–522.
3. K. M. v. Delft, J. C. T. Eijkel, D. Mijatovic, T. S. Druzhinina, H. Rathgen, N. R. Tas, A. van den Berg and F. Mugele, *Nano Lett.*, 2006, **7**, 345-350.
4. P. Mao and J. Han, in *Proceedings of the MicroTAS 2005 Conference*, Boston, MA, 2005, pp. 678-680.
5. G. Kittilsland, G. Stemme and B. Norden, *Sensors and Actuators A*, 1990, **21-23**, 904-907.
6. W.-H. Chu, R. Chin, T. Huen and M. Ferrari, *Journal of Microelectromechanical Systems*, 1999, **8**, 34-42.
7. S. W. Turner, A. M. Perez, A. Lopez and H. G. Craighead, *J. Vac. Sci. Technol., B*, 1998, **16**, 3835-3840.
8. H. Cao, Z. Yu, J. Wang, J. O. Tegenfeldt, R. H. Austin, E. Chen, W. Wu and S. Y. Chou, *Appl. Phys. Lett.*, 2002, **81**, 174-176.
9. C. P. Bean, M. V. Doyle and G. Entine, *Journal of Appled Physics* 1970, **41**, 1454.
10. J. A. Quin, J. L. Anderson, W. S. Ho and W. J. Petzny, *Biophys. J.*, 1972, **12**, 990.
11. R. I. Fleischer, H. W. Alter, S. C. Furman, P. B. Price and R. M. Walker, *Science*, 1972, **178**, 255.
12. H. Masuda and K. Fukuda, *Science (Washington, D. C., 1883-)*, 1995, **268**, 1466-1468.
13. S. Jeon, J.-U. Park, R. Cirelli, S. Yang, C. E. Heitzman, P. V. Braun, P. J. A. Kenis and J. A. Rogers, *Proc. Natl. Acad. Sci. U. S. A.*, 2004, **101**, 12428-12433.
14. D. J. Norris, E. G. Arlinghaus, L. Meng, R. Heiny and L. E. Scriven, *Advanced Materials*, 2004, **16**, 1393-1399.
15. Y. Zeng and D. J. Harrison, *Anal. Chem.*, 2007, **79**, 2289-2295.
16. B. R. Flachsbart, K. Wong, J. M. Iannacone, E. N. Abante, R. L. Vlach, P. A. Rauchfuss, P. W. Bohn, J. V. Sweedler and M. A. Shannon, *Lab Chip*, 2006, **6**, 667-674.
17. W. D. Volkmuth and R. H. Austin, *Nature (London)*, 1992, **358**, 600-602.
18. J. Han, J. Fu and R. B. Schoch, *Lab Chip*, 2008, **8**, 23-33.
19. J. Han, S. W. Turner and H. G. Craighead, *Phys. Rev. Lett.*, 1999, **83**, 1688-1691.
20. J. Han and H. G. Craighead, *Science*, 2000, **288**, 1026-1029.
21. R. E. Beck and J. S. Schultz, *Science*, 1970, **170**, 1302-1305.
22. A. G. Ogston, *Trans. Faraday Soc.*, 1958, **54**, 1754-1757.
23. J. Fu, P. Mao and J. Han, *Appl. Phys. Lett.*, 2005, **87**, 263902.
24. J. Fu, J. Yoo and J. Han, *Phys. Rev. Lett.*, 2006, **97**, 018103.
25. J. L. Viovy, *Rev. Mod. Phys.*, 2000, **72**, 813-872.

26. J. Fu, R. R. Schoch, A. L. Stevens, S. R. Tannenbaum and J. Han, *Nat. Nanotech.*, 2006, **2** 121 - 128.
27. S. W. P. Turner, M. Cabodi and H. G. Craighead, *Phys. Rev. Lett.*, 2002, **88**, 128013.
28. F. Tessier, J. Labrie and G. W. Slater, *Macromolecules*, 2002, **35**, 4791-4800.
29. M. Streek, F. Schmid, T. T. Duong and A. Ros, *J. Biotechnol.*, 2004, **112**, 79-89.
30. N. Laachi, C. Declet, C. Matson and K. D. Dorfman, *Phys. Rev. Lett.*, 2007, **98**, 098106.
31. Z. R. Li, G. R. Liu, Y. Z. Chen, J.-S. Wang, H. Bow, Y. Cheng and J. Han, *Electrophoresis*, 2008, **29**, 329–339.
32. D. Duong-Hong, J. Han, J.-S. Wang, N. G. Hadjiconstantinou, Y. Z. Chen and G.-R. Liu, *Electrophoresis*, 2008, **in review**.
33. D. Duong-Hong, J.-S. Wang, G. R. Liu, Y. Z. Chen, J. Han and N. G. Hadjiconstantinou, *Microfluidics and Nanofluidics*, 2007, **4**, 219-225.
34. Y.-C. Wang, A. L. Stevens and J. Han, *Anal. Chem.*, 2005, **77**, 4293-4299.
35. S. M. Kim, M. A. Burns and E. F. Hasselbrink, *Anal. Chem.*, 2006, **78**, 4779-4785.
36. J. H. Lee, S. Chung, S. J. Kim and J. Han, *Anal. Chem.*, 2007, **79**, 6868 -6873.
37. J. H. Lee, Y.-A. Song and J. Han, *Lab Chip*, 2008, **8**, 596 - 601.
38. S. J. Kim and J. Han, *Anal. Chem.*, 2008, **80**, 3507-3511.
39. R. B. Schoch and P. Renaud, *Appl. Phys. Lett.*, 2005, **86**, 253111.
40. R. B. Schoch, EPFL, Lausanne, Swiss, 2006.
41. R. S. Foote, J. Khandurina, S. C. Jacobson and J. M. Ramsey, *Anal. Chem.*, 2005, **77**, 57-63.
42. S. Song, A. K. Singh and B. J. Kirby, *Anal. Chem.*, 2004, **76**, 4589-4592.
43. A. V. Hatch, A. E. Herr, D. J. Throckmorton, J. S. Brennan and A. K. Singh, *Anal. Chem.*, 2006, **78**, 4976-4984.
44. A. E. Herr, A. V. Hatch, D. J. Throckmorton, H. M. Tran, J. S. Brennan, W. V. Giannobile and A. K. Singh, *Proc. Natl. Acad. Sci. U. S. A.*, 2007, **104**, 5268-5273.
45. Y.-C. Wang and J. Han, *Lab Chip*, 2008, **8**, 392-394.
46. J. H. Lee, Y.-A. Song, S. R. Tannenbaum and J. Han, *Anal. Chem.*, 2008, **80** 3198-3204.
47. R. F. Probstein, *Physicochemical Hydrodynamics : An Introduction*, Wiley-Interscience, 1994.
48. I. Rubinstein and B. Zaltzman, *Phys. Rev. E*, 2000, **62**, 2238-2251.
49. S. J. Kim, Y.-C. Wang, J. H. Lee, H. Jang and J. Han, *Phys. Rev. Lett.*, 2007, **99**, 044501.
50. O. B. Bakajin, T. A. J. Duke, C. F. Chou, S. S. Chan, R. H. Austin and E. C. Cox, *Phys. Rev. Lett.*, 1998, **80**, 2737-2740.
51. A. Balducci, P. Mao, J. Han and P. S. Doyle, *Macromolecules*, 2006, **39**, 6273-6281.
52. J. O. Tegenfeldt, C. Prinz, H. Cao, S. Chou, W. W. Reisner, R. Riehn, Y. M. Wang, E. C. Cox, J. C. Sturm, P. Silberzan and R. H. Austin, *Proc. Natl. Acad. Sci. U. S. A.*, 2004, **101**, 10979-10983.
53. W. Reisner, J. P. Beech, N. B. Larsen, H. Flyvbjerg, A. Kristensen and J. O. Tegenfeldt, *Phys. Rev. Lett.*, 2007, **99**, 058302.
54. Y. M. Wang, J. O. Tegenfeldt, W. Reisner, R. Riehn, X. J. Guan, L. Guo, I. Golding, E. C. Cox, J. Sturm and R. H. Austin, *Proc. Natl. Acad. Sci. U. S. A.*, 2005, **102**, 9796-9801.
55. R. Riehn, M. C. Lu, Y. M. Wang, S. F. Lim, E. C. Cox and R. H. Austin, *Proc. Natl. Acad. Sci. U. S. A.*, 2005, **102**, 10012-10016.
56. T. T. Perkins, D. E. Smith and S. Chu, *Science*, 1997, **276**, 2016-2021.

57. H. Cao, J. O. Tegenfeldt, R. H. Austin and S. Y. Chou, *Appl. Phys. Lett.*, 2002, **81**, 3058-3060.
58. G. C. Randall, S. K.M. and P. S. Doyle, *Lab Chip*, 2006, **6**, 516-525.
59. P. R. Nair and M. A. Alam, *Appl. Phys. Lett.*, 2006, **88**, 233120.233121-233123.
60. T. M. Squires, R. J. Messinger and S. R. Manalis, *Nat. Biotechnol.*, 2008, **26**, 417-426.
61. R. B. Schoch, L. F. Cheow and J. Han, *Nano Lett.*, 2007, **7**, 3895 -3900.
62. R. Karnik, K. Castelino, R. Fan, P. Yang and A. Majumdar, *Nano Lett.*, 2005, **5**, 1638-1642.

CHAPTER 3

Particle Transport in Micro and Nanostructured Arrays: Asymmetric Low Reynolds Number Flow

JASON PUCHELLA[1] and ROBERT AUSTIN[1]

[1] Department of Physics, Princeton University, Princeton, New Jersey, NJ 08544, USA

3.1 AN INTRODUCTION TO HYDRODYNAMICS AND PARTICLES MOVING IN FLOW FIELDS

This chapter is aimed specifically at providing an introduction to hydrodynamic flow fields in micro and nanofabricated arrays of obstacles. A far more complete introduction to nanofluidic flows can be found elsewhere in the excellent article by Squires and Quake.[1] While some of the material covered here by necessity is made redundant by Squires and Quake, the aim of this chapter is to bring out unexpected elements of flow in arrays not anticipated by Squires and Quake, and to make this understandable we will start from the beginning. Where we are unclear we urge the reader to consult Squires and Quake.

Everything begins with Newton's laws of motion. The force $\vec{F}$ acting on a mass element m is simply connected to its acceleration $\vec{a}$ by

$$\vec{F} = m\frac{d\vec{v}}{dt} \tag{3.1}$$

Unfortunately, the innocuous looking expression for the acceleration in Equation 3.1 is a total derivative which includes both space and time components. It is possible to write out explicitly the total derivative of the velocity in terms of the space and time derivatives. This derivation is not described in detail herein, however an excellent introduction (as always) can be found in Feynman's Lectures on Physics.[2] Written out as a function of the spatial and temporal derivatives, Newton's Law of motion in the absence of viscosity $\vec{F} = m\frac{d\vec{v}}{dt}$ for an incompressible fluid of density ρ becomes,

$$\rho[\vec{v}.\nabla]\vec{v} + \rho\frac{\partial\vec{v}}{\partial t} + \nabla P = -\rho\nabla\phi \tag{3.2}$$

where P is the local pressure and ϕ is an externally applied potential function, for example due to gravity. The space derivative $[\vec{v}.\nabla]\vec{v}$ known as the advective part of the total derivative, is perhaps counter intuitively a non-linear function of the velocity. It should be noted that '*advection*' is defined as the horizontal transport of matter by a velocity field $\vec{v}$, while '*convection*' is defined as the vertical transport of matter, and indirectly, a coupled thermal/density/gravitational effect. We will restrict ourselves to only using the word 'advection' in this chapter because of the confusion of convection with thermally driven processes that are coupled to gravity. However, we will discuss 3-D advective processes!

A little reflection can explain why the advective term is nonlinear in velocity: if we follow a differential volume *PdV* in space as it is advected by flow, the total change in the velocity vector of fixed scalar magnitude due to transport around an obstacle is the product of the magnitude of the scalar value multiplied by the magnitude of the spatial derivative of $\vec{v}$.In the simple case of pure circular motion with a fixed radius of curvature *R* we find that the magnitude of the advective derivative is given by $\vec{a}_{advect} = -\frac{v^2}{R}\hat{r}$, which is nothing more than the centripetal acceleration $|a_{centripetal}|$of a point particle travelling in a circle of radius *R* with speed v, as every freshman physics student knows. Of course, $|a_{centripetal}|$is in fact non-linear in *v* as promised. Thus, these formidable vector calculus equations actually have an intuitive meaning if one is able to see through the mathematics. The advective acceleration term $\rho[\vec{v}.\nabla]\vec{v}$, which is the density times the spatial acceleration of velocity field, will later be called (confusingly!) the inertial force, rather like confusingly calling the centripetal acceleration the "centrifugal force". The inertial force is no more a force than the centrifugal force is, but the term is hopelessly embedded in the literature and fighting it makes one look like a pedant.

Since Equation 3.2 is a nonlinear partial differential equation, there is no obvious closed form solution to this equation, and hence even "dry water", to use Feynman's evocative phrase, can consume a lifetime of study easily, and indeed has for some people! However, in spite of its complexities it does have one thing going for it: since there is no viscosity there is no loss of kinetic energy and one can write down a potential function whose gradient is the velocity. But this is of little help in the real world since we have to add viscosity to have any hope of entering the nanofluidic world where shear forces become extremely important. Viscous drag originates from transverse shear of the velocity and the subsequent transfer of particle momentum between layers. When viscosity is added to the Equation 3.2 the Navier-Stokes (N-S) equation is obtained:

$$\rho[\vec{v}.\nabla]\vec{v} + \frac{\partial \vec{v}}{\partial t} + \nabla p - \eta \nabla^2 \vec{v} = -\rho \nabla \phi \quad (3.3)$$

Since viscous drag is dissipative it results in the "sucking" of kinetic energy from a volume element in shear and results in non-energy conserving flows. Thus, the curl of the velocity $\nabla \times \vec{v}$ is not necessarily zero in the presence of viscosity flow and one can no longer write a potential function from which all the velocities can be computed at subsequent time, as you can for dry water. Since the Navier-Stokes equation has the double misfortune of being both nonlinear in $\vec{v}$ and also not energy conserving it is an extraordinarily difficult physics problem for arbitrary initial values of $\vec{v}$, ρ and η to calculate subsequent values of, $\vec{v}$ quite hopeless.

Fortunately (perhaps) some simplifications in the Navier-Stokes equation are possible. The spatial derivatives in the Navier-Stokes Equation are numerically of the form,

$$\nabla \cdot \vec{v} \sim \frac{|\vec{v}|}{L} \quad (3.4)$$

where L is some characteristic length scale over which the velocity changes "significantly" in direction or magnitude. In the case of a nanofabricated structure, for example a post with radius R, the parameter L would be the radius of the post, and the spatial derivative is due to the curving of the fluid around the post. The magnitude of the viscous drag term in Equation 3.3 is,

$$|\eta\nabla^2\vec{\upsilon}| \sim \frac{\eta|\vec{\upsilon}|}{L^2} \quad (3.5)$$

while the advective derivative has magnitude,

$$|\rho[\vec{\upsilon}\cdot\nabla]\vec{\upsilon}| \sim \frac{\rho\upsilon^2}{L} \quad (3.6)$$

The ratio of the advective derivative to the viscous derivative is given by the Reynolds number, R_e, i.e.

$$R_e \sim \frac{|\rho[\vec{\upsilon}\cdot\nabla]\vec{\upsilon}|}{|\eta\nabla^2\vec{\upsilon}|} \sim \frac{\rho\upsilon L}{\eta} \quad (3.7)$$

The usual statement is that if $R_e << 1$, as is demonstrably true for water flowing ($\eta = 0.01$ poise for water at 20°C) in structures with characteristic length scales for velocity changes on the order of $L \sim 10^{-4}$cm or less, one can completely ignore the advective derivative and assume that the equation governing the fluid flow is given by,

$$\frac{\partial\vec{\upsilon}}{\partial t} + \nabla P - \eta\nabla^2\vec{\upsilon} \sim 0 \quad (3.8)$$

which is again an equation linear in $\vec{\upsilon}$. But it is important to warn the reader that symmetry plays a role in these equations through the spatial derivatives. Thus, even if locally the viscous drag term $\eta\nabla^2\vec{\upsilon}$ dominates the advective term $[\vec{\upsilon}\cdot\nabla]\vec{\upsilon}$ in the Navier-Stokes equation, if the advective term by reason of the symmetry of the nanostructure has a consistent (small) value which sums with movement through an array of obstacles, it could in principle result in a significant displacement of the volume element even at low R_e. For example, there is of course a finite advective derivative to a particle even at very low R_e as it goes around a post (the advective term is the centripetal acceleration, which is certainly not zero), and this finite term can move an object across streamlines even at very low R_e. In the case of the "bump" arrays we will discuss later, movement across streamlines even at very low R_e is critical in understanding how the "bump" array works.

Diffusion also can compromise the transport of particles in arrays of obstacles even at very low R_e if structures are sufficiently small so that the times to diffuse a distance L are comparable to the time it takes for the particle to be transported that same distance by the streamline velocity. The significance of particle diffusion is indicated by the Péclet number (P_e) for mass diffusion,

$$P_e = \frac{L\upsilon}{D} \quad (3.9)$$

where L is a characteristic length of structure, υ is the local velocity, and D is the mass diffusivity. The Péclet number is easily derived: the time t_υ to be carried in a streamline a distance L is simply $t_\upsilon = L/\upsilon$ while the time t_D to diffuse a distance L is $t_D = L^2/2D$. Hence the ratio of the time to diffuse to the time to be advected by a streamline $P_e = L\upsilon/D$. The balance between diffusive transport and advective transport can lead to some non-intuitive behaviour. Mass diffusion is a dissipative effect and results in no net force perpendicular to streamline flow. That is, in the absence of other coupled effects, diffusion will not result in a shift in the mean location of a particle distribution. However, steric repulsion and advective acceleration at each post adds a non-linearity into transport.

3.2 POTENTIAL FUNCTIONS IN LOW REYNOLDS NUMBER FLOW

The problem faced is calculating $\vec{\upsilon}$ in large areas filled with obstacles. That is, we have a terrible boundary value problem to solve. Even the N-S equation with the advection term $\rho[\vec{\upsilon}\cdot\nabla]\vec{\upsilon}$ removed is NOT easy to solve for arbitrary boundaries to the flow field since the N-S equation is not a scalar, but instead is a vector differential equation. However, there are two simplifications that we can make if all of the complicated shape boundaries are in the *x-y* plane, but in the *z* axis there is a simple straight etch (vertical walls). The two simplifications are that the *z*-depth etch *h* is effectively infinite compared to the spaces between the walls in the *x-y* plane (i.e. deep etch), or that the *z*-depth of the structure *h* is much less than the *x-y* dimensions of the problem (i.e. shallow etch). In either case the velocity function can be separated into functions $v = (x, y)$ and *f(z)*:

$$\upsilon(x, y, z) \sim v(x, y) \times \vec{f}(z) \tag{3.10}$$

Although this may not seem like much of a simplification (it is not a trivial separation of variables), it allows us to write a scalar stream function $\psi = (x, y)$ which connects the boundary values of the velocity flow to the velocity field within a complex array of objects. Presumably we know what *υ(z)* is for the infinite depth and very shallow etch depths. The problem then becomes finding $v(x, y)$ using some connection between $\psi(x, y)$ and $v(x, y)$. We will spare the reader the ugly details (they can be found in references [3] and [4]) and will simply state some important results. If we let

$$v(x, y) = \nabla \times \psi(x, y)\hat{z} \tag{3.11}$$

and also let $\psi(x, y) = \psi(x, y)\hat{z}$ for convenience, remembering that the scalar function $\psi = (x, y)$ is the heart of $\vec{\psi}$. The low R_e N-S equation then becomes,

$$\eta\nabla^2[\nabla \times \vec{\psi}(z)] = \nabla P \tag{3.12}$$

Some vector calculus then yields that for the very deep etch structure (*h* >>> *x,y* for all x,y in the structure),

$$\nabla^4\psi = 0 \tag{3.13}$$

which is known as the biharmonic equation and is a standard result found in all good hydrodynamic text books.[4] In reality, most nanotechnology devices are not extremely deeply etched, in fact they are more typically extremely shallow ($h << x,y$ for all x,y in the structure), and the approximations used to obtain the biharmonic equation (Equation 3.13) are not valid. In the case of a very shallow etch a different equation can be derived for the stream function. First, for the very shallow etch we know that the z dependence of the velocity function must be a parabolic solution of the 1-D low R_e N-S equation:

$$f(z) \sim v_0 \left[1 - \left(\frac{2z}{h}\right)^2\right] \tag{3.14}$$

In this case we can write Equation 3.8 as,

$$\nabla^2 \vec{\text{v}}(\text{x}, \text{y}) \cdot f(\text{z}) + \frac{2}{\text{h}^2} \vec{\text{v}}(\text{x}, \text{y}) = \nabla \text{P}(\text{x}, \text{y}) \tag{3.15}$$

We then use some vector calculus identities to get the Laplacian partial differential equation for the stream function in the case of very shallow etches, i.e.

$$\nabla^2 \psi(\text{x}, \text{y}) = 0 \tag{3.16}$$

Thus, the stream function $\psi(x, y)$ for a shallow etched nanofluidic system satisfies Laplace's equation, and hence solving the stream function for a thin flow, low R_e N-S equation is effectively an electrostatic calculation driven by the boundary values, a huge simplification! If a solution for the stream function $\psi(x, y)$ can be found, the velocity in the *x-y* plane can be found from Equation 3.11. The pressure field can be found by solving, unfortunately, the following equation:

$$-\frac{h^2}{8\eta} \nabla P = \nabla \times \vec{\psi} \tag{3.17}$$

If we had not insisted on explicitly making the velocity in the thin plane a function of both x and y, this equation may have been written much more simply if the flow is only in the x direction, that is, effectively a 1-D flow where the z-axis etch depth is very shallow. In this case one obtains the simple equation of lubrication theory,[5]

$$\frac{\partial P}{\partial x} = \frac{\partial^2 \upsilon}{\partial z^2} \tag{3.18}$$

Before nanofabrication came around, this equation was adequate to explain simple thin laminar flows. However, as we will show by using nanofabrication techniques it is quite possible to make thin flows with complex (and useful) changes in the x and y directions that cannot be handled using Equation 3.18. We are therefore forced to develop a more encompassing model for our problem of an array of obstacles in the *x-y* plane.

In fact, using the techniques of nanofabrication it is easily possible to create thin structures which, while they remain in the thin flow limit used to derive Equation 3.17, have varying etch depths. Such techniques have been exploited by the Craighead group at Cornell University in variable depth arrays.[6] When the etch depth becomes a function *h(z)* Equation 3.17 must be changed to the rather formidable form:

$$\nabla^2\vec{\psi} = -2\nabla(\ln d(z)) \times (\nabla \times \vec{\psi}) \tag{3.19}$$

In principle Equation 3.19 can be used to solve for the streamlines of an arbitrarily complex array of obstacles with an etch depth which varies in height, but the mathematics become substantial, and are strongly coupled with the difficulties of setting the boundary values correctly. An example of large area variable etch depth nanofluidic flow can be found in the work by Darnton *et al.*,[7] where 3 layers of fluid were stacked on top of each other across a wide channel to provide rapid mixing for large fluidic volumes. The flow profiles calculated by using Equation 3.19 are shown in Figure 3.1, and are taken from reference [7].

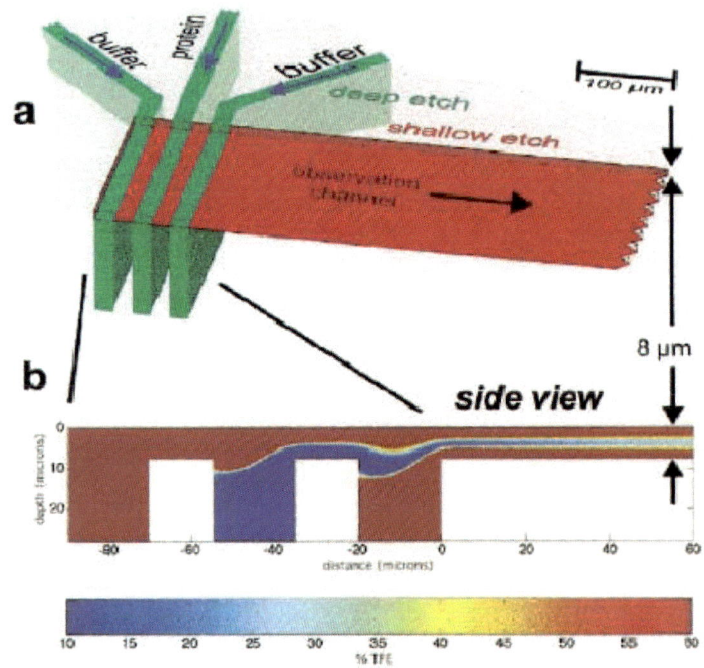

Figure 3.1 Design of a 3-D mixing chip. (a) Top view of the chip. The 80-micron-deep inlet channels are shown in green, and the 8-micron-deep outlet channel is shown in red. (b) Two-dimensional fluid dynamics simulation with false colour representation of the TFE concentration. A jet of the centre (protein) solution between two layers of buffer solution is formed. The TFE concentration is computed by adding the advective term to the diffusion equation. Adapted from reference [7] an reproduced with permission.

There are some simple and powerful ways to use the thin etch Laplace's Equation 3.16 however. It was previously mentioned that the stream function basically solves a charge-free electrostatic problem. In electrostatics such problems can be solved by specifying the Dirichlet boundary condition (setting the value ψ needs to take at the boundary) or the Neumann boundary condition (setting the value of the derivative of ψ at the boundary).[8] Since $\nabla \times \vec{\psi} = v(x,y)$ it is clear that Neumann boundary conditions can be used to set the velocity fields within a nanofluidic enclosed volume. In practice, this means that the boundary current flow (which sets the boundary value of incoming fluid velocity) can be held fixed if the current is injected from a high impedance current source: that means a high pressure source of fluid should be run through a long thin tube which acts as a high impedance fluidic resistor. This "current source" then sets the Neumann boundary condition and one can back-propagate from the fixed current flow and calculate the fluid velocities everywhere in principle within a closed area, if there are no major obstacles within the area.

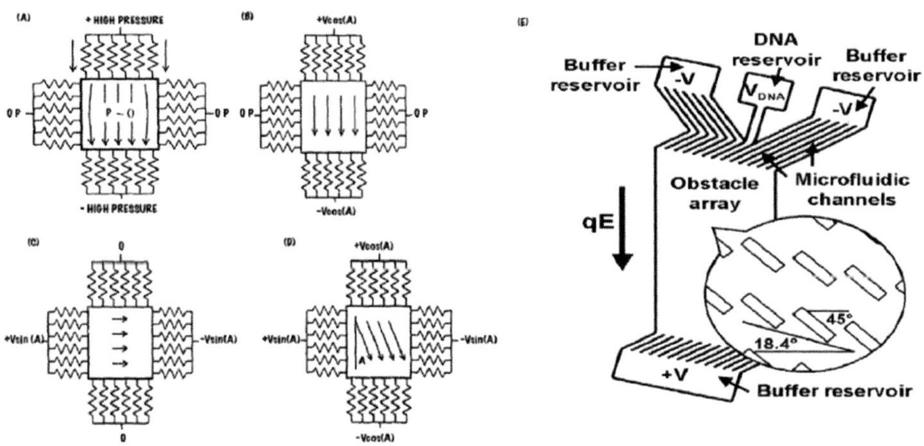

Figure 3.2 (A) The basic idea behind controlling flow patterns at low *Re*: wall arrays of current injectors. (B-D) In this case, the fluid is forced to uniformly flow at a constant angle θ. (E) A more complex design to inject a narrow jet of fluid in an array of tilted obstacles.

Figure 3.2 shows a simple example of setting Neumann boundary conditions for an open region to achieve a flow field tilted at an angle *A* to the sides of the open area. Figure 3.3(A) shows a corner of an etched chip, while Figure 3.3(B) shows the remarkably stable laminar flows that can be propagated across a large area using this concept of a wall of injectors.

3.3 ARRAYS OF OBSTACLES AND HOW PARTICLES MOVE IN THEM: PUZZLES AND PARADOXES IN LOW R_E FLOW.

Up to this point we have only considered the movement of a fluid in an open area nanofabricated device. However, we have not considered the transport of particles or the effects of complex obstacle patterns within a nanofluidic device. At very low Reynolds numbers the effect of boundary conditions and particle size can critically perturb particle trajectories from the streamlines that can be calculated using the techniques outlined in the above section. While some of these effects (such as steric displacement) scale effectively with size there are various microscopic phenomena that couple into the transport behaviour to drastically alter net transport of particles.

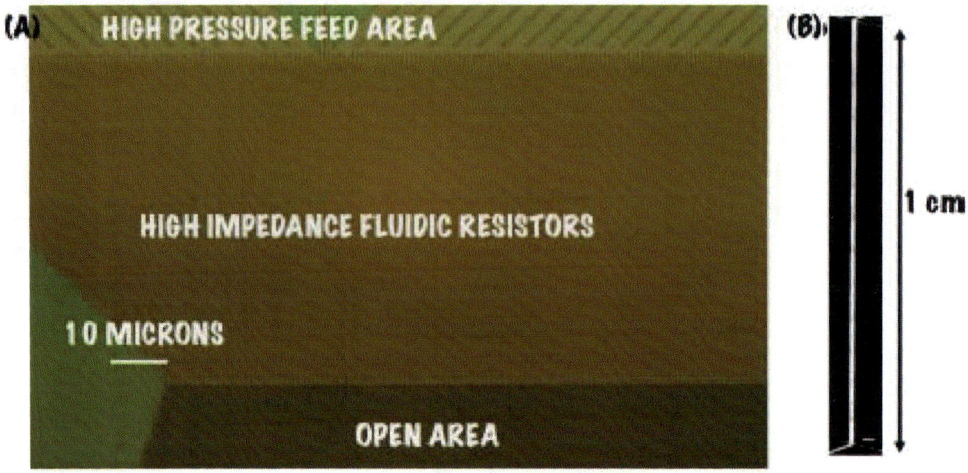

Figure 3.3 (A) Corner of an etched device consisting of a wall of current injectors. (B) Laminar jet of fluorescent 1.0 micron balls propagating through an open area, with flow determined by the boundary conditions.

Surprisingly, once particulate transport comes into play the so-called "inertial" term in the N-S comes into play, even at low R_e. Thus, although the vast majority of work in nanofluidics is performed under conditions that lead to several simplifications, totally ignoring what are called the inertial terms can be a major mistake. When one has complex boundaries and particles in transport some of the simplifications we have discussed above become cloudy in their validity. For example, while it is true that in steady-state, incompressible flow $\delta \upsilon / \delta t = 0$, this does not mean that the fluid acceleration at any point in the volume is exactly zero. This time independent acceleration is expressed in the N-S equation as $\mathrm{v} \cdot \nabla \mathrm{v}$. This term can be rewritten as a combination of translational and rotational acceleration:

$$\mathrm{v} \cdot \nabla \mathrm{v} = 1/2 \nabla \mathrm{v}^2 + \nabla \times \mathrm{v} \times \mathrm{v} \tag{3.20}$$

Thus, the inertial term $\mathrm{v} \cdot \nabla \mathrm{v}$ can give rise to a finite curl ($\nabla \times \mathrm{v}$) to the velocity field if the fluid turns a corner. This curl term can give rise to a net circulation of the fluid, and this circulation can give rise to unusual trapping of particulates in the flow field even at relatively low R_e. The classic example of this effect is the flow of a fluid in a pipe of radius a as it rounds a bend of radius R, a condition called Dean Flow.[9] As the liquid rounds the bend the curl induced by the rotating fluid gives rise to a circulation of two vortices which rotate in opposite directions. The stream function ψ that we mentioned earlier (Equation 3.11) has the form,

$$\Psi = D^2 f(r) \sin(\theta) \tag{3.21}$$

where the Dean number D is given by,

$$D = R_e \left[\frac{a^{1/2}}{R} \right] \tag{3.22}$$

Although the amplitude of the vortex is very small at low R_e, it is a collective rotation and can integrate with distance if the flow is set in a spiral. Because of this circulation and the subsequent stall lines particles can be trapped in the radial plane of the flow, and this effect has substantial use in microfluidics.[10,11]

We will now concentrate on the "bump arrays" that were invented in our group at Princeton University.[12] Figure 3.4 shows a schematic of such a device and also describes the basic theory of the transport process in a bump array. A microfluidic channel is filled with a matrix (or array) of microfabricated obstacles. Each row of obstacles is shifted horizontally with respect to the previous row by $\varepsilon\lambda$, where λ is the centre-to-centre distance between the obstacles. To simplify the discussion, let us assume that ε equals 1/3. Particles to be separated are driven through the matrix by a laminar fluid flow. Because of the low Reynolds number in these devices, the flow lines are laminar and deterministic. That is to say, there is no turbulence, and all inertial effects are presumably negligible (but maybe not). Particles whose diameter D is large compared to the slot width set by the gap between the posts g times ε will not follow individual streamlines, but instead be propelled by many streamlines. This fundamentally changes their final migration direction. A particle that is too big to fit into slot width $\beta\varepsilon g$ is physically displaced so that its hydrodynamic centre-of-mass is once again in slot 2. This process is repeated every time a large particle approaches a row of obstacles, and the net result is the bumping of large particles and the downward transport of small particles deterministically. If the bump array is not tilted the net effect would purely be to enhance diffusive mixing without changing the location of the centre of the particle distribution.

However, since the array is in fact tilted, the next post encounter is biased: particles that moved to one side of the stall line will be advected so that they approach the next post in a similar fashion (along a stall line), while particles that moved to the other side will be advected away from the stall line. In this way, the process repeats from post-to-post. In effect, diffusive motion is biased away from the array tilt direction.

The solid lines drawn in Figure 3.4 represent streamlines that make boundaries between the zig-zaging bundles of laminar flow. These lines are interesting in that they are also stall lines: although they represent finite velocities of flow, they are not continuous but rather apparently *terminate* at the surfaces of the posts. The number of rows between these stall point terminations is $1/\varepsilon$. Since the fluid flow is incompressible $(\nabla \cdot \vec{\upsilon} = 0)$, in the vicinity of the post along the line the liquid must come to a rest and hence these lines terminate at places where the fluid velocity is instantaneously zero.

In Figure 3.4 we assume that there is a parabolic fluid profile between the posts in the x-y plane of the posts, which is a crude simplification that comes from the deep-etch analysis given in Equation 3.13 for parallel walls, not circles. However, crude as this approximation may be it does lead to an interesting insight: the bundles of current flux which are divided into quanta of $1/\varepsilon$ bundles in an array of tilt ε are not of equal width, because of the parabolic fluid profile: bundles near the walls are bigger than those in the centre.

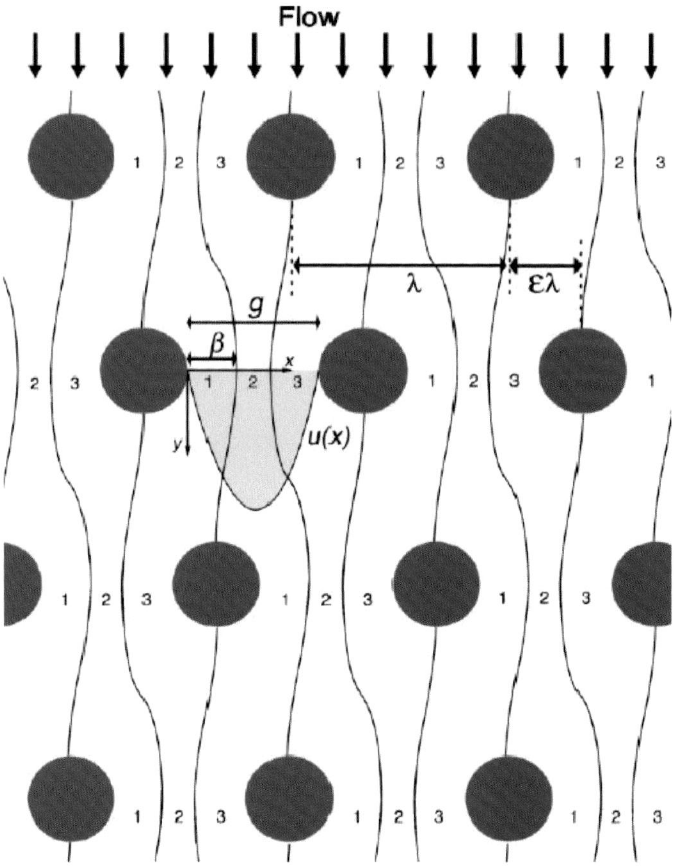

Figure 3.4 Top view diagram of streamlines in low Reynolds number flow through an array of posts. Each row is shifted to the right by one third of the post-to-post spacing, λ, making the row shift fraction $\varepsilon = 0.33$. Three equivalent streamlines flow between each gap, numbered 1 to 3, which cyclically permute from row to row. The streamlines are divided by stall lines which begin and terminate at the posts. An example flow profile $u(x)$ is drawn into one of the gap regions. β is the width of the first streamline. Adapted from reference [13] and reproduced with permission.

A detailed calculation of this first order approximation modelling real flows can be found in reference [13]. The result of this analysis is shown in Figure 3.5. These data show that although one would expect that as $\varepsilon \Rightarrow 0$ the critical bumping radius D_c would go to zero, in fact it remains finite and never gets below approximately 0.2 g: so to separate nano-objects you need a nano-array!

Even the analysis of reference [13] is an approximation. The stall points are places of zero velocity where the decision to *bump* or *not bump* around a post is made. At these stall points where particle instantaneous velocities are zero the local Péclet number tends to zero and diffusional fluctuations in advective velocity and particle size variations can blur the bumping transition. The problem is that the very presence of the particle itself strongly perturbs these stall lines at the critical point where the decision to "*bump or not bump around a post*" has to be made, and analytical approaches we fear fail because of the dramatic consequences of small perturbations.

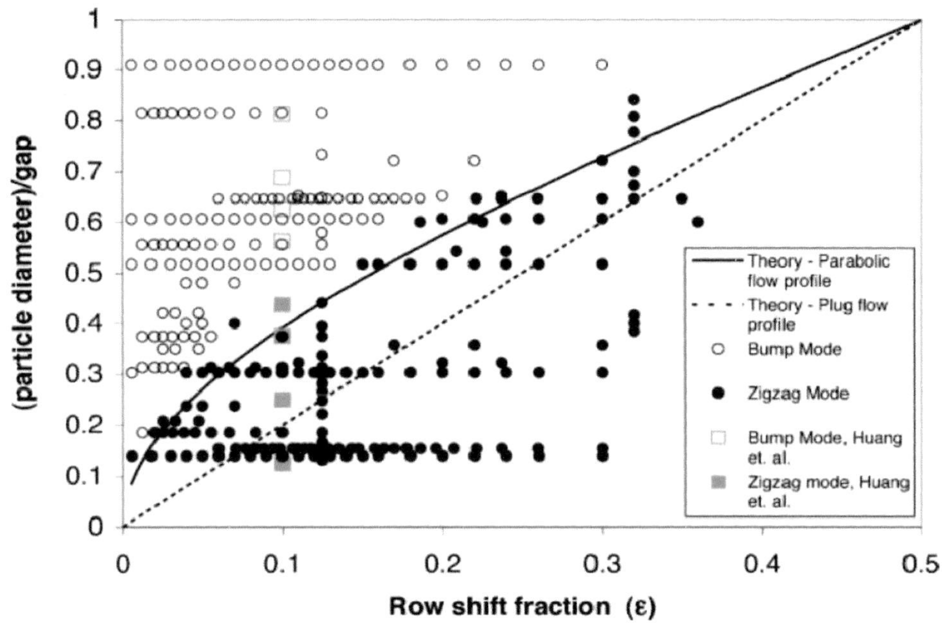

Figure 3.5 Experimental points of the particle diameter divided by the gap, versus the row shift fraction, ε. For this work (in black) and that of Huang *et al.* (in grey), open points represent bump mode and solid points represent zigzag mode. Zigzag mode particles follow the streamlines, while bump mode particles follow the array slope, ε.

A feeling for the sensitivity of the motion of particulates to local variations in the stream flow can be seen by simply examining the velocity field in an array of posts in the vicinity of the stall points. At the stall point analytical techniques fail and it is necessary to go to hydrodynamic code simulations, although even those fail when a particle is present! We will not go into this complex subject here, nor will we attempt to discuss the harder problem of addressing the number of steps required to sufficiently resolve the space between particles and posts and the number of post interactions required to accrue a significant bump displacement in a computational simulation.

We present, to get some flavour for the problem, a full N-S simulation of the fluid flows around the posts in a bump array using the software package COMSOL.[14] There are three figures, which present firstly the velocities and pressure profiles (Figure 3.6) and secondly the forces moving the particles (Figure 3.7 and Figure 3.8).

There is always a danger when working with simulations that intuition gets thrown out the window and with that a real understanding of the physics of the problem. However we find these simulations most illuminating. To illustrate some of the fluid dynamic behaviour in a bump array, we chose a two-dimensional geometry based on reference [12] with ε set to 0.1. However, the methods presented here should be general since the full N-S equation is used. A two-dimensional simulation should be a good approximation over much of the channel, although aspects such as the Dean flow we discussed above will not be seen, and as the etch depth increases the validity of the two-dimensional approximations to the true three-dimensional flow will become questionable.

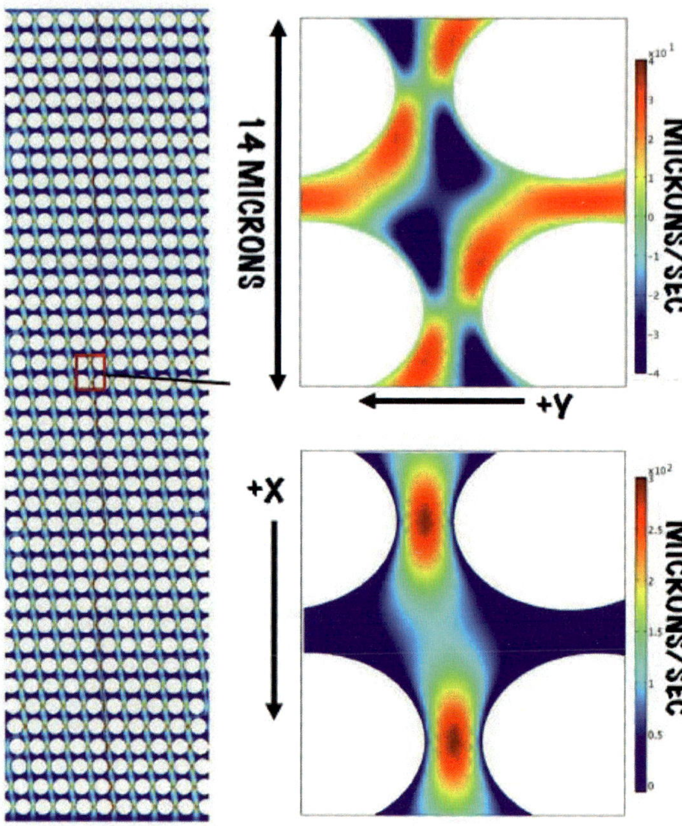

Figure 3.6 Velocity fields within a typical bump array driven by hydrostatic pressure. Left: Velocity fields demonstrate a characteristic pattern over the array where boundary conditions determine local deviations near the walls. Upon closer inspection (right), the velocity field is seen to be dominated by flow along the channel (x-velocity) modified by slower but dynamically very significant displacement perpendicular to the flow direction (y-velocity).

All posts and side walls are treated as no-slip surfaces, which also may not be strictly true under true nanofluidic conditions.[15] The fluid (water of density 1 g/m and viscosity of 0.001 Pa.s) input was maintained at a constant flow rate of 1000 microns/s normal to the wall on the left side (thus representing fairly high rates of flow in the system). The output (right hand wall) was maintained at zero pressure with no viscous stress. It should be noted that this means that the fluid need not exit the channel with a velocity normal to the exit wall. This is analogous to watering flowing from a pipe that has a directional baffle before the end that deflects flow upon exiting. Moreover, in Figure 3.6 the bifurcation in the y velocity that occurs at the midpoint of the gap g indicating the sensitive nature of this point to minor perturbations and diffusional broadening should be noted. Since it is the y velocities that are responsible for the 'bumping' process (movement to the right in spite of the vertical walls) the interplay between diffusional broadening and advective transport become extremely critical at this gap midpoint point, and as one shrinks the scale to the nano regime the interplay becomes ever more critical.

A surprising aspect of the forces that determine bumping and the competition with diffusion can be seen in Figure 3.7 and Figure 3.8 which show the force densities (extracted from the COMSOL simulation of the full NS equation) acting on the water elements as they move through the structure. The viscous damping forces are dissipative and always act in the opposite direction from the local velocity, while the "inertial" forces are sensitive to the gradients in the velocity and thus can locally change sign as the gradients flip direction. This effect can be seen in Figure 3.8(B). The gap g is shown, and a bundle of streamlines that pass out of the gap to the left represent the "reset" streamline bundle βg shown in Figure 3.4.

We can estimate the size of various forces we expect in this system based on some simple arguments. The viscous force density is $f_{vis} \approx \frac{\eta \upsilon}{D^2} = 10^6$ N/m^3. Here, the characteristic size $D \sim 10^{-6}$ m corresponds to the gap between posts. Note that $f_p \approx f_{vis}$ as expected for viscid flow. These are to be compared to the inertial force density $f_{iner} \approx \frac{\rho \upsilon^2}{D} =$ 10^3 N/m^3. That is, the forces involved with moving the fluid around posts are about three orders of magnitude smaller than the pressure and inertial forces. It should also be noted that the inertial force density is dominated by the acceleration and de-acceleration of the fluid elements in the direction of the motion and that the centripetal acceleration, which points along the radial direction of the circular motion around the post, is not visible in Figure 3.8. However, while a purely laminar low R_e flow in a plane would be expected to have zero curl in the velocity stream, the full N-S equation through the inertial terms does in fact allows non-zero curl even at low R_e, as was demonstrated in Equation 3.20. Of course, our COMSOL simulation also yields non-zero vorticity for the flow at those regions of the flow field where the centripetal acceleration is greatest, as can be seen in Figure 3.8(C). These vortices will rotate objects that are embedded in the flow, which means that non-spherical objects can have quite different flow properties in micro/nanoarrays then simple spherical ones, although calculation of these dynamics given the complexities of the flow even at low R_e can be quite daunting.

We close our discussion with an illustration of the complexities of low R_e flow in complex, asymmetrical structures. One of the great "mysteries" in the bump array work was hidden in the original publication.[12] Figure 3.9(A) shows the remarkable separation that was achieved using the ideas outlined in this review. However, Figure 3.9(B) shows a strange phenomenon: as the flow rate (and hence the Péclet number) decreases, we would expect the jet profiles of the bumped particles to broaden due to diffusion, and the sharp edge of the angle versus particle diameter to broaden on both sides of the critical diameter. But, this is not what happens. Particles with diameters greater than the critical diameter, D_c continue to bump, while those with a diameter less than the critical diameter do start to move at smaller angles. However the width of the streams does not appreciably grow as you would expect to happen in diffusional broadening. That is, as the Péclet number decreases the dispersion of the device changes, becoming less binary with decreasing speed. However the resolving power of the device remains roughly the same. A small random movement across a stall line will mean that a particle of sufficient size will undergo a steric displacement to either one side of the streamline or the other that will be unrecoverable (on average) by diffusion when the displacement is sufficiently large. Clearly this effect is particle size, time and velocity dependent in complex ways.

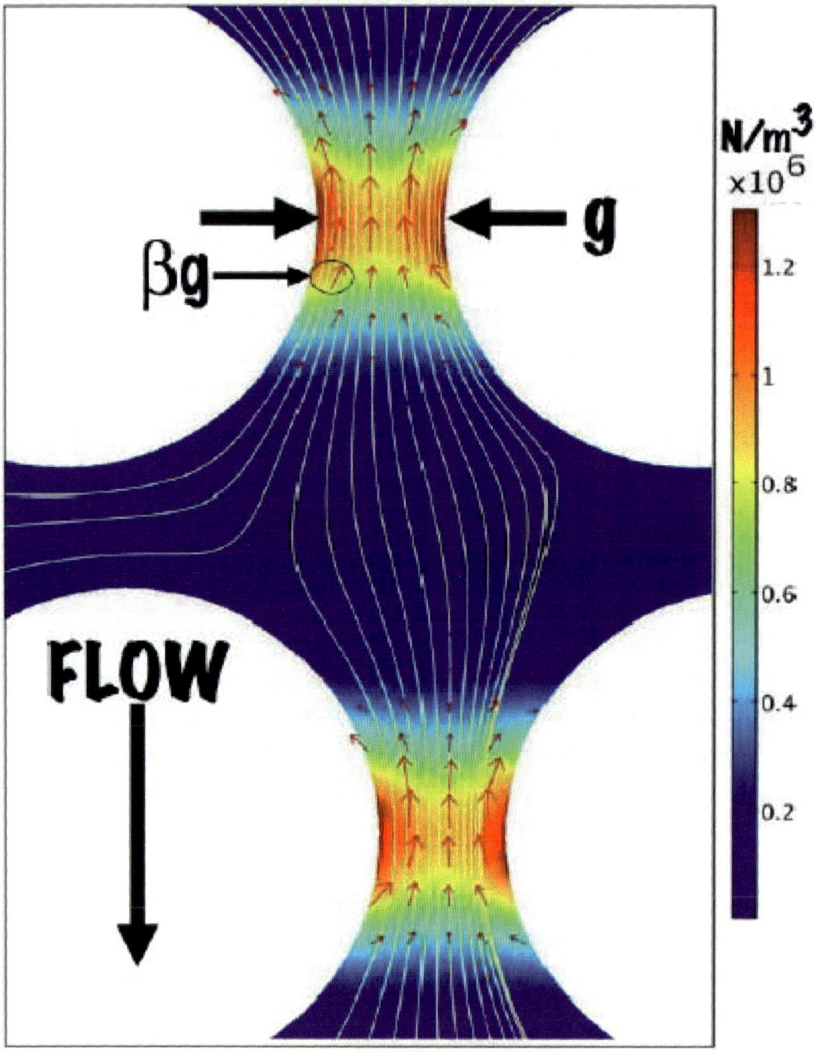

Figure 3.7 The viscous drag forces/volume acting on the fluidic elements under hydrostatic pressure head. The gap g used in Figure 3.4 is shown, and the bundle of streamlines βg that is shunted to the right is also shown. The field is 14 microns long.

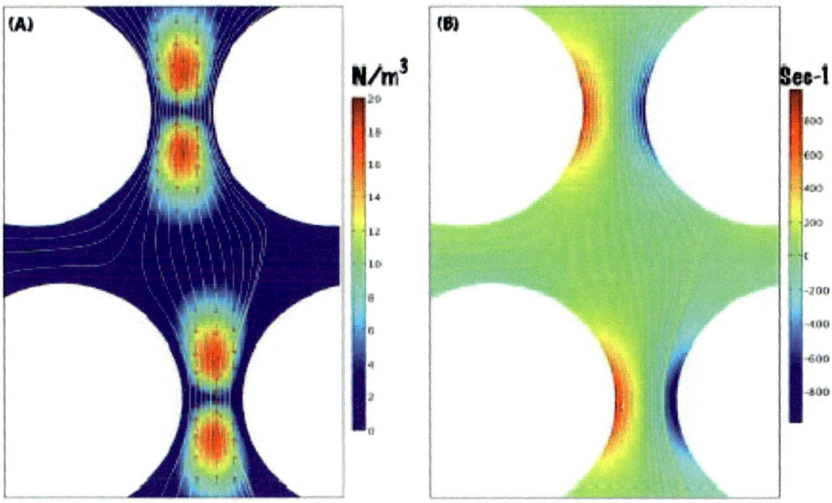

Figure 3.8 (A) The inertial terms in the N-S equation. Note the vastly different change in magnitude of the inertial terms for this simulation. (B) The vorticity $\omega = \nabla \times \upsilon$ for the flow. The field is 14 microns long.

We do not understand very well at this point the strange behaviour of the particles in this device, although we can make some intuitive guesses as to what ‘may’ be happening. The bump decision occurs in the gap *g* shown in Figure 3.5. Particles near the left side of gap *g* that are below the critical diameter D_c move with the “reset” streamline bundle βg, while particles greater than the critical diameter move to the right, at high Péclet numbers. However, diffusional broadening presumably moved some particles into the “reset” bundle that normally would travel to the left as a bumped particle, resulting (again presumably) in a broadened jet. However, because diffusional distances only increase at the $t^{1/2}$, diffusion outside the gap will have little influence on bump jet broadening, and within the gap the diffusion is bounded: diffusion to the left is restricted by the presence of the wall, while diffusion to the right will result in bumping. Hence, because of the broken symmetry of the bump array the diffusional bias is towards maintaining the free-channel bumping even for sub-critical diameter particles at low R_e and perhaps the maintenance of the jet stream width. But the issue is certainly made more complex by the blockage of the flow by the particles themselves. Low R_e laminar flow may be simpler than the complexities of high R_e turbulent flow, but its application to micro and nanofluidics reveals that there are still many tricks to be learned, and many surprises awaiting those of us pushing the boundaries of nanofludics.

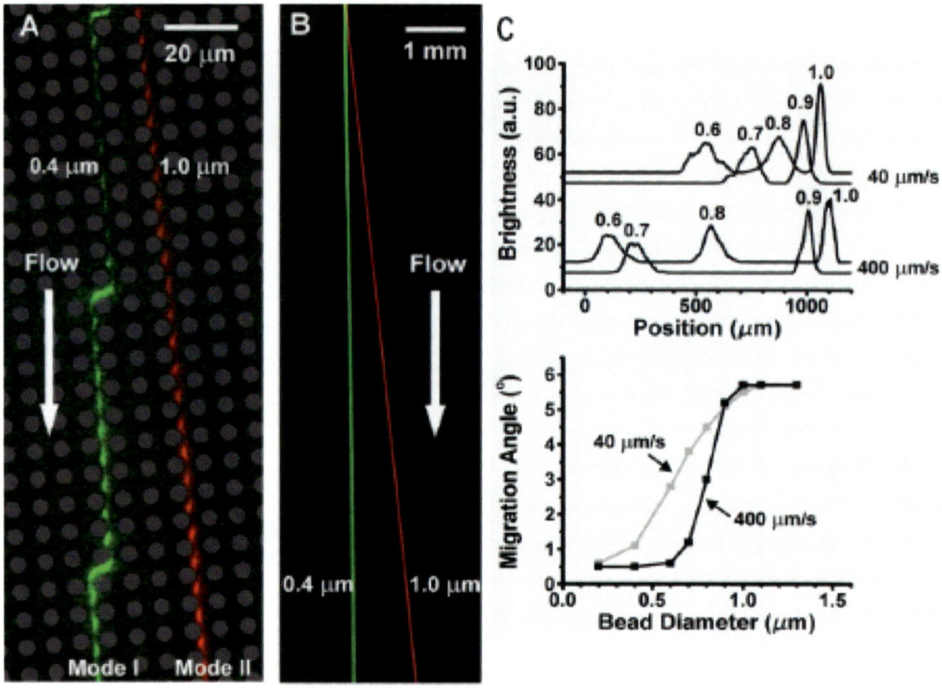

Figure 3.9 Fluorescence images of microspheres migrating in the obstacle matrix, showing (A) the two transport modes for (B) the separation of microspheres with no dispersion. The grey dots in (A), which represent the obstacles, have been superimposed on the fluorescent image. (C) Fluorescent profiles of microspheres separated using flow speeds of 40 microns/s (upper curves) and 400 microns/s (lower curves) scanned at 11 mm from the injection point. The 0.60 micron, 0.80 micron, and 1.03 micron diameter beads are green fluorescent, while the 0.70 micron and 0.90 micron are red, and thus each scan is shown as two curves representing the two colours. The measured migration angles as a function of microsphere diameter at two different flow speeds are shown in the lower part of (C).

References

1. T.M. Squires and S.R. Quake, *Rev. Mod. Phys.,* 2005, **77**, 9771026.
2. R.P. Feynman, R.B. Leighton and M. Sands, *Lectures in Physics*, 2006, Vol. II Chapter 40, Addison-Wesley, Menlo Park.
3. N. Darnton, Ph.D. Thesis: Protein Folding and 2.5 D Hydrodynamics, Princeton University Dept. of Physics, 2002.
4. V.N. Constantinescu, *Laminar Viscous Flow*, 1995, Springer-Verlag, New York.
5. R. L. Panton, *Incompressible Flow (3rd ed.)*, 2005, Wiley, New York.
6. J. Han and H. G. Craighead, *Science*, 2000, **288**, 1026.
7. E. Kauffmann, N.C. Darnton, R. H. Austin, C. Batt, and K. Gerwert, *Proc. Natl. Acad. Sci. USA*, 2001, **98**, 6646-6649.
8. J.D. Jackson, *Classical Electrodynamics, 3rd ed.*, 1999, Wiley, New York.

9. M. Johnson and R.D. Kamm, *J. Fluid Mech.*, 1986, **172**, 329-345.
10. D. Di Carlo, D. Irimia, R.G. Tompkins, and M. Toner, *Proc. Natl Acad. Sci. USA,*2007, **104**, 18892-18897.
11. C.J. Tsai, and D.Y.H. Pui, *Aerosol Science and Technology*, 1990, **12**, 813-831.
12. L.T. Huang, E.C. Cox, R.H. Austin, and J.C. Sturm, *Science*, 2004, **304**, 987-990.
13. D.W. Inglis, J.A. Davis, R.H. Austin, and J.C. Sturm, *Lab Chip*, 2006, **6**, 655-658.
14. COMSOL, Inc., 1 New England Executive Park, Suite 350, Burlington, MA 01803
15. J.K. Holt, H.G. Park, Y.M. Wang, M. Stadermann, A.B. Artyukhin, C.P. Grigoropoulos, A. Noy, and O. Bakajin, *Science*,2006, **312**, 1034-1037.

CHAPTER 4

Molecular Transport and Fluidic Manipulation in Three Dimensional Integrated Nanofluidic Networks

T.L. KING[1,4], X. JIN[2,3], N. ALURU[2,3] & P.W. BOHN[4,5]

1 Department of Chemistry, University of Illinois at Urbana-Champaign, Urbana, IL 61801, USA.
2 Department of Mechanical Science and Engineering, University of Illinois at Urbana-Champaign, Urbana, IL 61801, USA.
3 Beckman Institute for Advanced Science and Technology, University of Illinois at Urbana-Champaign, Urbana, IL 61801, USA.
4 Department of Chemical and Biomolecular Engineering, University of Notre Dame, Notre Dame, IN 46556, USA.
5 Department of Chemistry and Biochemistry, University of Notre Dame, Notre Dame, IN 46556, USA.

4.1 INTRODUCTION

The development of micro total analysis systems (μ-TAS) is driven by the desire for faster, lower cost analysis of complex sample mixtures in a single, compact device. The ability to carry out multiple operations in a single, integrated architecture reduces sample consumption and loss (in terms of mass and concentration), thereby enabling the facile characterization of those samples that are either endogenously (e.g. neurotransmitters and pheromones) or exogenously (e.g. toxins) limited in mass. The advantages of microfluidic-based analytical systems, including high analytical throughput, small sample volumes, reduced materials costs and minimal power consumption, are clear, but miniaturization of fluidic systems presents real problems for injection, fluidic isolation and mixing, especially when these systems are constrained to a two-dimensional format.

The three-dimensional integration of microfluidic systems utilizing nanofluidic elements to couple vertically separated microfluidic channels creates hybrid, three-dimensional architectures with demonstrated advantages in several problematic areas. These include maintenance of distinct chemical environments,[1] digital manipulation of fluid aliquots within and between these environments[2-4] and rapid mixing of fluids.[5,6] Since nanofluidic elements are crucial to accomplishing these objectives, this chapter will address recent theoretical and experimental contributions aimed at understanding fluid flow in nanofluidic systems, as well as those factors unique to hybrid microfluidic-nanofluidic architectures. The application of these hybrid systems to chemical analysis will also be discussed.

Nanofluidic technology has enabled further miniaturization and enhanced functionality for μ-TAS applications and impacted many other areas as well. Research into

DNA characterization in nanostructures has been particularly prolific,[7] including the separation of DNA and other biomolecules based on both entropic trapping and Ogston sieving[8-10]. The similar sizes of both synthetic and biological nanopores have allowed powerful detection modalities for biomolecules[11] and synthetic nanoparticles[12] to be developed. Moreover, the unique transport properties characteristic of nanofluidic structures have allowed the development of devices for drug delivery,[13] fluidic transistors,[14] separation of ionic species,[15,16] sample desalting prior to electrospray mass spectrometry,[17] studies of surface adsorption[18] and constrained biomolecules dynamics,[19] as well as many other applications.

The size scale of nanofluidic systems relative to microscale and larger systems engenders several significant differences. Geometric scaling rates cause the surface-area-to-volume ratio of nanofluidic systems to be significantly larger than that of microscale systems, increasing the influence of surface chemistry on system behaviour. This is especially relevant when considering the impact of surface charge density on nanofluidic flow. A single 50 nm (diameter) x 5 µm long cylindrical pore with a surface charge density of $\sigma = 2 \times 10^{-3}$ C/m^2 has approximately 1000 immobile surface charges. If this pore is filled with a 1.6 mM solution of a 1:1 electrolyte, there are just enough counterions to maintain electroneutrality in the pore, effectively creating a system in which all ions inside the pore are counter ions, with co-ions being almost totally excluded.[20] Further lowering of the electrolyte concentration can create a situation where the ion concentration inside the pore is greater that the bulk ion concentration.[21] In addition, nanofluidic channels can have sizes on the order of the Debye length (*vide infra*) allowing access to transport phenomena not observed at larger scales. For example, the Debye length, κ^{-1}, ranges from 1 nm to 10 nm for univalent electrolyte solutions of ionic strength, μ, between 100 mM and 1 mM. Additionally, diffusion becomes a viable mass transport mechanism on the nanoscale simply due to the small distances involved. This eliminates the myriad problems experienced within low Reynolds number (laminar flow) regimes in microfluidic systems. For example, a molecule with a diffusion constant of $D = 10^{-6}$ cm^2/s travels approximately 100 nm in 100 µs, thus establishing the characteristic dimensions and timescales over which chemical communication can occur by diffusion. Finally, the characteristic dimensions of nanofluidic elements and of large molecular structures (e.g. biopolymers) can be similar, enabling unique separation and detection modalities such as stochastic sensing with biological and synthetic nanopores.[22-24]

At the same time that they present exciting new avenues for research, the size scales and novel phenomena encountered in nanofluidic systems present significant experimental challenges. The synergistic coupling of theory and experiment allows nanofluidic phenomena to be understood at a fundamental chemical level that would be inaccessible by experiment alone. The seminal 1965 work of Rice and Whitehead[25] provides a framework for understanding recent experimental efforts. In this study, they considered an infinitely long, cylindrical capillary of radius a, filled with a univalent electrolyte solution. For small zeta potentials ($\psi_o < \sim 50$ mV), the potential experienced in solution, ψ, at a distance, r, from the pore axis can be calculated from the Poisson-Boltzmann equation,

$$\frac{1}{r}\frac{d}{dr}\left(r\frac{d\psi}{dr}\right) = \kappa^2\psi \tag{4.1}$$

where the inverse Debye length, κ, is given by,

$$\kappa = \left(\frac{8\pi n e^2}{\varepsilon k T} \right)^{\frac{1}{2}} \tag{4.2}$$

Here, n is the ion number density, ε is the dielectric constant, k is the Boltzmann constant, and T is the temperature. Knowing that ψ is finite at $r = 0$ and that $\psi = \psi_o$ when $r = a$ allows the potential to be calculated. Combining this with the Poisson equation gives

$$\rho(r) = -\frac{\varepsilon \kappa^2}{4\pi} \psi_0 \frac{I_0(\kappa r)}{I_0(\kappa a)} \tag{4.3}$$

where $\rho(r)$ is the net charge density and I_o is a zeroth-order Bessel function of the first kind. This allows an equation of motion in the presence of both an applied pressure gradient and an axial electric field to be written, i.e.

$$\frac{d^2 v_z}{dr^2} + \frac{1}{r}\frac{dv_z}{dr} = -\frac{P_z}{\eta} - \frac{E_z}{\eta}\rho(r) \tag{4.4}$$

Here v_z is the axial velocity, P_z is the applied axial pressure gradient, η is the viscosity, and E_z is the applied axial electric field. Upon substituting the net charge density $\rho(r)$ and applying appropriate boundary conditions, the solution to Equation 4.4 is given by

$$v_z(r) = \frac{P_z}{4\eta}(a^2 - r^2) - \frac{\varepsilon \psi_0}{4\pi\eta}\left[1 - \frac{I_0(\kappa r)}{I_0(\kappa a)}\right] \tag{4.5}$$

Under conditions where the applied pressure is zero, i.e. when only an electrical potential is applied, Equation 4.5 reduces to

$$v_z(r) = -\Omega E_z \left[1 - \frac{I_0(\kappa r)}{I_0(\kappa a)}\right] \tag{4.6}$$

where Ω is comprised of the constants in the second (electrokinetic) term of Equation 4.5. Clearly, for values where $\kappa a >> 1$ Equation 4.6 reduces to

$$v_z(r) = -\Omega E_z \tag{4.7}$$

demonstrating that for large capillary diameters and/or low Debye lengths, κ^{-1} (high ionic strengths), ion migration is the dominant electrokinetic transport phenomenon, and plug flow is recovered. Conversely, for small capillary diameters and/or large Debye lengths (low ionic strengths), i.e. conditions where $\kappa a \leq 1$, electroosmosis dominates transport. This important theoretical contribution presents experimentalists with an array of handles to manipulate fluid flow in nanofluidic systems, since surface charge density, ionic strength, characteristic nanostructure dimension, and applied electrical potential all can be used to control the manner in which the κa dimensionless group couples to nanofluidic transport. Below we describe the efforts of experimentalists to verify these theoretical predictions and exploit them for practical utility.

4.2 EXPERIMENTAL CHARACTERIZATION OF NANOFLUIDIC FLOW

4.2.1 Surface Charge

The large surface area-to-volume ratio of nanofluidic channels allows fixed surface charges to exert a large influence on flow behaviour. A clear demonstration of this is contained in work by Chun and Stroeve, who measured the diffusive transport of two charged proteins through nanoporous polycarbonate track etched (PCTE) membranes (with a diameter of 10 nm) modified with charged alkanethiol surface assembled monolayers (SAMs).[26] PCTE membranes were coated with Au by electroless deposition[27] and mercaptoundecanoic acid ($HS(CH_2)_{10}COOH$) monolayers were assembled on the pore surfaces. The diffusive fluxes of bovine serum albumin (BSA, pI = 4.7) and bovine hemoglobin (BHb, pI = 7.0) were then measured as a function of pH. The transmembrane diffusion of each protein was maximal at their respective pIs, an observation attributed to the minimization of electrostatic repulsion between the negatively charged pore walls and the proteins (Figure 4.1). Interestingly, the flux of both proteins is reduced at pH values below their pI as well as above. The authors hypothesize that while the proteins are attracted to the nanopore in this regime due to their positive charge, the electrostatic exclusion of associated negative counterions results in diminished fluxes. Analogous results were obtained for diffusive transport of tyrosine and phenylalanine,[28] and diffusion of a series of lectins was also observed to exhibit the same behaviour.[29] Further experiments were performed in which the membrane surface charge was controlled by applying electrical potentials between the Au nanocapillary array membrane (NCAM) and a Ag/AgCl reference. These experiments show that the diffusive flux for highly charged species (BHb at pH 4.7, $z \approx 14$) is reduced in rough proportion to the magnitude of the applied potential, while fluxes of less charged species (BSA at pH 4.7, $z \approx 2$) are unaffected,[30] providing further evidence for the role of surface charge in controlling fluid flow on the nanoscale.

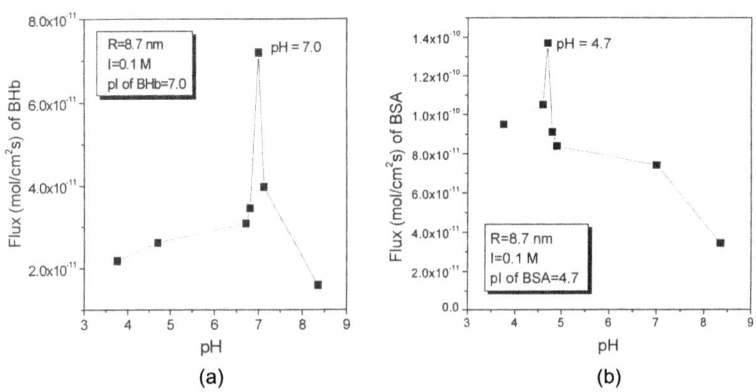

Figure 4.1 Flux of bovine hemoglobin (BHb) (*a*) and bovine serum albumin (BSA) (*b*) across a PCTE/Au/mercaptodecanoic acid membrane as a function of solution pH. Adapted from Chun, K.Y. and P. Stroeve, *Protein transport in nanoporous membranes modified with self-assembled monolayers of functionalized thiols.* Langmuir, 2002, **18**, 4653 and reproduced with permission.

Electrostatic exclusion and transport selectivity occur in nanocapillary systems for both large macromolecules and small ionic species. For example, Lee and Martin imparted switchable ion permselectivity to Au NCAMs (pore radius = 1.4 nm) by chemisorption of cysteine to the NCAM surface.[16] At pH values above the cysteine isoelectric point (pKa ≈ 6) the membranes exhibit preferential transport of cations, rejecting anions, with the reverse being true for pHs less than the pI of cysteine. Additionally, under comparable conditions small ions are transported more rapidly than large ions in this system due to hindered transport. Similar size exclusion effects have been demonstrated for F^-/Cl^- separations in layer-by-layer (LBL) constructed multilayer polyelectrolyte nanofiltration membranes.[15]

Martin and co-workers observed that ion rectification, i.e. fluidic diode-like behaviour, occurred in single conical Au nanopores (d_{base} = 600 nm, d_{tip} = 10 nm, l = 12 μm) fabricated on poly(ethylene terephthalate) membranes with single damage tracks[31] only when a surface charge was imparted to the pore[32]. For example, when the pores are filled with 0.1 M KF, current versus voltage curves are Ohmic. However, the same experiment carried out in 0.1 M KCl produces current versus voltage curves that exhibit much larger currents ("*on*" behaviour) at negative potentials than at positive potentials ("*off*" behaviour) of the same magnitude. The rectifying behaviour is attributed to the absorption of Cl^- at the gold surface imparting a negative charge, which, because of the comparable sizes of the electrical double layer and the pore tip, generates an electrostatic trap for cations near the narrow tip of the pore when positive potentials are applied, leading to the observed ion rectification. This trap is not formed when negative potentials are applied resulting in larger currents.

4.2.2 Debye Length

The examples above have focussed mainly on the contributions of surface charge. However, it is impossible to explain nanofluidic flow properly without a detailed understanding of the effects of the electrical double layer. In fact, the very classification of a system as 'nanofluidic' depends on the size of the nanofluidic structures relative to the Debye length.[33] It is important to consider the effects of the Debye length as coupled to effects of surface charge, since the Debye length controls the magnitude of the electrostatic potential experienced by ions at different radial positions inside a nanocapillary. For example, in a microchannel or at high ionic strengths in a nanochannel ($\kappa a >> 1$) the electric potential generated by immobile surface charges rapidly decays to the bulk solution value. However, at low ionic strengths ($\kappa a \approx 1$) the electric potential even in the centre of a nanochannel can deviate from the bulk potential (Figure 4.2).[14] To maintain intra-pore electroneutrality, nanochannels can become a unipolar solution of counterions, balancing the immobile surface charge, even to the extent of becoming enriched in counterions relative to the bulk.[21] This effect was elegantly demonstrated by Stein, Kruithof, and Dekker[34] in a silica slit-pore device, where at low ionic strengths conductance depends only on surface charge density and is independent of slit height or salt concentration.

It should be noted that the evidence above for characteristic nanofluidic phenomena influenced by fixed surface charges occurred in solutions where the Debye length occupied a significant portion on the nanochannel(s). In fact, by simply reducing the ionic strength from 0.1 M to 0.01 M, thus increasing the Debye length relative to the nanopore size, Ku and Stroeve increased separation selectivities for BSA/BHb transport from ~ 7 to 67, a 10-fold increase.[35]

In an elegant series of experiments Karnik and co-workers[36] measured ionic conductance through silica slit pores (120 μm long, 3.5 μm wide and 30 nm high) across a range of ionic strengths before and after biomolecular adsorption to the nanochannel surface, thereby demonstrating the relationship between surface charge, Debye length, pore size and fluid flow behaviour. Upon adsorption of aminosilane or biotin, conductance increases in the low ionic strength regime (below 0.01 M) due to increased surface charge density, while conductance at higher ionic strengths is identical to the bulk. Adsorption of much larger species, such as streptavidin (ca. 5 nm globular diameter), also increases conductivity at low ionic strengths while lowering conductivities at higher ionic strengths, due to steric reduction in nanochannel diameter. Analogous results have been obtained by Schoch and co-workers in Pyrex slit-pores.[37-39]

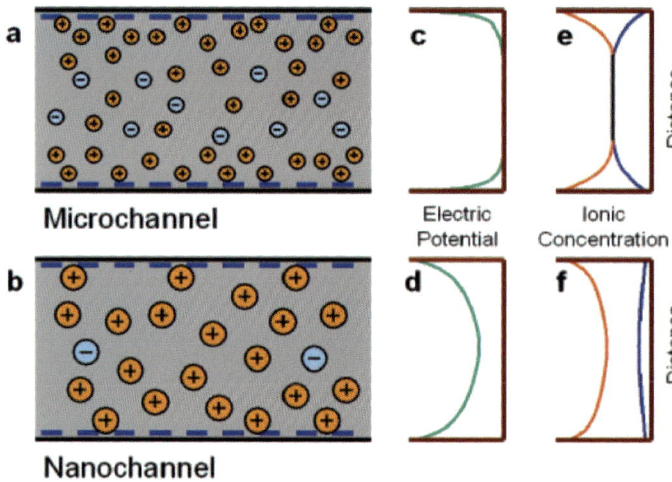

Figure 4.2 Effect of the Debye length on electric potentials and ion populations. (a) When $\kappa a >> 1$ bulk and micro/nanochannel ion concentrations are nearly identical. (b) When $\kappa a \approx 1$ the solution inside a nanochannel must become enriched in ions of one charge to maintain electroneutrality. (c) The electric potential rapidly decays to the bulk value when $\kappa a >> 1$. (d) When $\kappa a \approx 1$ the electrical potential can deviate from the bulk due to the influence of immobile surface charges. (e) The concentration of cations (orange) and anions (blue) when $\kappa a >> 1$ equals the bulk concentration. (f) The concentration of counterions (orange) is enriched relative to coions (blue) when $\kappa a \approx 1$. Adapted from Karnik, R., *et al.*, *Electrostatic Control of Ions and Molecules in Nanofluidic Transistors.* Nano Letters, 2005, **5**, 943 and reproduced with permission.

The rate of ionic diffusion across conical nanopores has been shown to depend on the direction of a concentration gradient relative to the geometric polarity of the asymmetrical pores.[40] Concentration gradients were established across conically track-etched poly (ethylene terephthalate) nanopores with $r_{base} \approx 300$ μm and 1.5 nm $< r_{tip} <$ 20 nm by placing 1.0 M and 0.1 M KCl solutions on alternate sides of the membrane. When the 0.1 M solution contacts the nanopore tip ($c_{tip} = 0.1$ M) and the 1.0 M solution contacts the base ($c_{base} = 1.0$ M), currents due to ionic diffusion are greater than when the solutions are reversed. This can be understood in terms of the relative sizes of the Debye length and r_{tip} in the two cases. When $c_{tip} < c_{base}$ interaction between the surface charge and ions in solution is greater than when $c_{base} < c_{tip}$ due to the longer Debye length in the narrower tip region. This leads to greater permselectivity in favour of K^+ over Cl^-, as evidenced by larger currents.

Further evidence that surface charge effects on transport in nanofluidic systems are mediated by the relative sizes of the Debye length and the nanofluidic channel is given by Kemry *et al.*[41] and by Kuo and co-workers[42] in their studies of both diffusive and potential-driven transport across PCTE membranes. In this system the magnitude and sign of the nanopore surface charge control the rate and direction of nanofluidic transport. More significantly, for 15 nm PCTE membranes which exhibit a positive zeta potential, the dominant mechanism of electrokinetic transport changes from electroosmosis at low ionic strengths (μ = 10 mM) to ion migration at high ionic strengths (μ = 1 M). Thus, for the same applied electrical potential, charged species flow in opposite directions depending on the relative sizes of the nanopores and of the Debye length. Exploiting the same idea, Garcia, *et al.* switched the elution order of a (negatively charged/neutral) pair of probes, by varying nanostructure diameter at constant ionic strength.[33]

4.3 INTEGRATED NANOFLUIDIC SYSTEMS

Clearly, flow in nanofluidic systems is dominated by the influence of fixed surface charges on ions in solution, as measured by the relative size of the Debye length (and thus the ionic strength of the solution). These results demonstrate that the magnitude, direction of transfer, and permeability of nanofluidic membranes can be controlled by external factors such as applied electrical potential, ionic strength, pH, nanostructure dimension, and molecular adsorption. Significant recent activity has focused on exploiting these phenomena for the creation of integrated fluidic systems. Specifically, nanofluidic elements have been used to create sophisticated, three-dimensionally integrated architectures for the digital manipulation, isolation, and pre-concentration of fluidic volume elements (voxels).

4.3.1 Molecular Sampling (Digital Fluidic Manipulation)

The ability to manipulate (individually address and transfer) small fluid voxels represents an important enabling capability for the development of μTAS. Kuo and co-workers first demonstrated this digital fluidic manipulation capacity in three-dimensional hybrid microfluidic/nanofluidic devices consisting of a polycarbonate NCAM functionalized with poly(vinylpyrrolidone) (PVP) sandwiched between two perpendicular, vertically separated microfluidic channels.[3,43] The NCAM provides a fluidic connection between the microchannels that is impermeable when no external potential is applied ("*off*" state). This can be externally actuated by applying an electrical potential to the system ("*on*" state). For example, applying a positive potential, $\Delta E = E_{receiving} - E_{source}$, across a 15 nm NCAM between a receiving microchannel containing 5 mM pH 8 phosphate buffer and a source channel filled with the same buffer as well as 170 nM fluorescein causes an aliquot of fluorescein solution to be transported from the source to the receiving channel (Figure 4.3(a)).[43] If an identical experiment is performed with a 200 nM NCAM, a negative ΔE is required to translocate fluorescein from the source to the receiving channel. This reversal in net flow direction is explained by the competition between the oppositely directed electroosmotic flow (EOF) vectors due to the negatively charged PDMS microchannels ($\zeta < 0$) and the positively charged PVP moieties immobilized on the NCAM surfaces ($\zeta > 0$). For large (200 nm) diameter NCAMs most of the electrical potential (98%) is dropped across the microchannels and thus the direction of the EOF vector in the microchannels determines net transport. However, in 15 nm NCAMs ~25% of the potential is dropped across the nanocapillaries (Figure 4.3(b)), and the higher resultant electric field allows the NCAM-based EOF to determine the transport direction.[3]

The behaviour of these hybrid systems has been exploited to carry out a variety of analytical unit operations necessary to realize a complete μTAS device. Kuo and co-workers[3] showed that above a condition-dependent threshold voltage, mass transfer during electrokinetically-actuated injections across an NCAM from one microchannel to another can be made 100% efficient, i.e. every analyte molecule in the microfluidic region above the NCAM is transferred. This observation enables a particularly facile and repeatable method of sample injection for on-chip electrophoresis, as demonstrated by Cannon *et al.* in gated sample injections of a mixture of amino acids followed by electrophoretic separation, with separation reproducibilities of 1%.[2] Gated analyte injections have also been demonstrated for Pb^{2+} solutions into DNAzyme filled channels[44] and for organo-mercaptans, followed by adsorption to Au colloids.[45] Conversely, Tulock and co-workers[4] showed that it is possible to select a specific band from an electrokinetic separation and transfer it across an NCAM to a separate fluidic environment for further processing, establishing a preparative separation capability at the attomole level. Electrokinetic injections across NCAMs into PDMS electrophoresis channels have also been used for sample preparation prior to electrophoretic separations.[46]

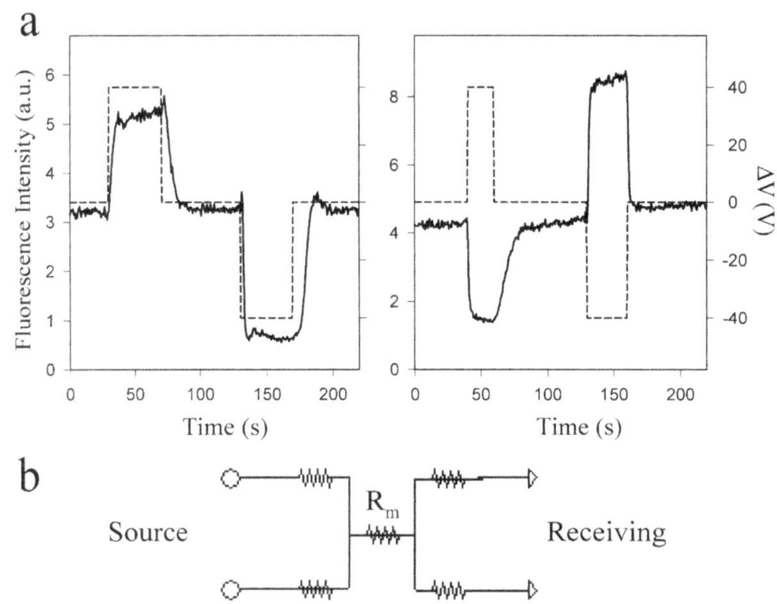

Figure 4.3 (a) Transport of fluorescein solutions across (*left*) 15 nm and (*right*) 200 nm PCTE nanocapillary arrays connecting two 100 μm PDMS microchannels as a function of applied potential and time. (b) Impedance network model of a hybrid microfluidic/nanofluidic device. Adapted from Kuo, T.C., *et al.*, *Gateable nanofluidic interconnects for multilayered microfluidic separation systems.* Anal Chem, 2003, **75,** 1861 and reproduced with permission.

Electrokinetic injections across NCAMs in hybrid microfluidic/nanofluidic systems have also been shown to effect rapid mixing of the transported fluidic elements with the solution in the receiving channel.[6] Rapid mixing adjacent to the NCAM is largely a function of the small intra-pore spacing in an NCAM, typically a few hundred nm. A molecule with a diffusion coefficient of 5×10^{-7} cm^2/s can diffusively access these distances on the μs-ms timescale. Rapid mixing is also observed in depth, as well as laterally, most likely due to convection resulting from the non-laminar character of fluid translocation across an NCAM.

A separate and wholly novel series of nanofluidic systems based on immobilized networks of vesicles and lipid nanotubes has been developed by Orwar and co-workers.[5,47-51] Pressure driven[49] and electrokinetic[48] transport of single nanoparticles and externally created vesicles[51] has been demonstrated in these systems. The small size and biological verisimilitude of these systems naturally lend themselves to application in the study of confined biomolecular reactions[5] and especially to the study of membrane bound proteins.[47]

4.3.2 Sample Pre-Concentration

The practical necessities of fluid manipulation external to integrated micro- or nanofluidic systems (i.e. processing samples and loading fluidic devices) require minimum sample volumes on the order of μL. Because many samples are either intrinsically or extrinsically mass-limited, achieving these fluid volumes often involves significant dilution, making analyte pre-concentration a necessity. One of the first examples of pre-concentration using nanofluidic elements integrated in a microfluidic device is provided by Khandurina and co-workers.[52] A silicate plug was formed between two co-planar microfluidic segments, allowing the passage of ions but not a DNA PCR marker when an electrical potential was applied between them, resulting in a ~100-fold concentration of the DNA marker in approximately 5 minutes. A similar approach from the same laboratory led to a 600-fold concentration of proteins in 8 minutes.[53] Similarly, application of electrical potentials across a hydrogel plug immobilized within a planar microfluidic channel resulted in 500-fold concentration of DNA as well as a 50-fold concentration of fluorescein, a much smaller molecule.[54] Clearly, permselectivity is the root cause of the pre-concentration phenomena observed in these experiments. The permselectivity of NCAMs separating microfluidic channels has been most stringently tested by Fa *et al.*[1] in their studies of H^+ isolation. Spatiotemporal maps of microchannel proton concentration were used demonstrate the capability of NCAMs with positive ζ-potential to exclude trans-membrane transport of hydrogen ions. Under conditions of double-layer overlap, $\geq10^3$-fold H^+ concentration differentials can be maintained for periods of hours, even while sustaining EOF in the microchannel adjacent to the NCAM.

Pre-concentration of analyte molecules in three-dimensional hybrid microfluidic-nanofluidic devices was first observed by Kuo *et al.*[3] and was extended by Zhang and Timperman.[55] Kuo observed 3-fold enhancement in the concentration of fluorescein above a 15 nm NCAM sandwiched between two microfluidic channels when a trans-NCAM electrical potential was applied.[3] Zhang created devices in a similar configuration with a 10 nm, positively-charged NCAM and observed that while anionic analytes such as fluorescein or FITC-labelled peptides were concentrated, neutral and positive (e.g. rhodamine 6G) analytes pass through readily.[55] Size-based concentration of large molecules such as DNA has also been demonstrated in an identical geometry.[56]

Concentration factors in the millions have been achieved for proteins in a system designed by Wang *et al.*[57] that illustrates both the fundamental physical causes of pre-concentration as well as the interplay between micro- and nanofluidic elements in hybrid systems (Figure 4.4). In-plane arrays of nanochannels of height ≤ 50 nm were used to fluidically connect two microchannels. When a potential, E_n, is applied across the nanochannels, counterions (in this case cations due to the negative surface charge of the nanostructure) are preferentially transported. As E_n is increased, the transport of ions across the nanochannels is diffusion limited due to the formation of an ion-depleted region in the microchannel. This ion depletion causes a larger Debye length, increasing the charge selectivity of transport, thereby creating a positive feedback mechanism for ion-depletion region formation. Above some threshold value of E_n a layer in the microchannel totally devoid of co-ions is created, essentially an extension of the nanochannel Debye layer filling a region of the microchannel. Application of a potential across the microchannel (normal to E_n) then drives analyte molecules into the space-charge region interface via electroosmosis of the second kind, causing an accumulation of co-ions (in this case negatively charged ions). Further evidence for nonlinear electrokinetic effects in this system, including fluidic vortices within the depletion region has also been reported.[58] A nanofluidic concentrator based on these principles has also been created in an all PDMS microfluidic device, with nanofluidic gap formation between two microchannels via electrical junction breakdown.[59]

It should be noted that specific chemical recognition elements can be incorporated into nanofluidic structures to create systems that constitute *de facto* pre-concentrators due to the efficiency of reaction in spatially constrained environments as well as the innate localization of captured analyte molecules in the nanofluidic membrane. Antibody-antigen interactions have been used for protein pre-concentration by immobilizing antibody fragments to electroless Au plated NCAMs,[19] and Ni^{2+}-His[60] or Cu^{2+}-His[61] interactions have been used to pre-concentrate proteins by modifying surface-grafted polymer brushes with organometallic compounds.

4.4 THEORY AND SIMULATIONS

Clearly a number of exciting applications are made possible by the unique transport properties of nanofluidic structures. These are enabled by the complex interplay of electrodynamics, fluid physics, and chemistry at small dimensions, but if we are to continue to exploit these unusual characteristics for future applications, we must develop a sound first-principles-based understanding of fluid physics and chemistry at nanometer length scales. In this section we investigate ion enrichment-depletion phenomena at micro-nanofluidic junction regions and the unique current signatures resulting from them by studying the ionic concentration and potential variation in coupled micro- and nanochannels and the resulting convective flow. Together these basic physicochemical events constitute the basis for the very powerful fluidic manipulation strategies described above.

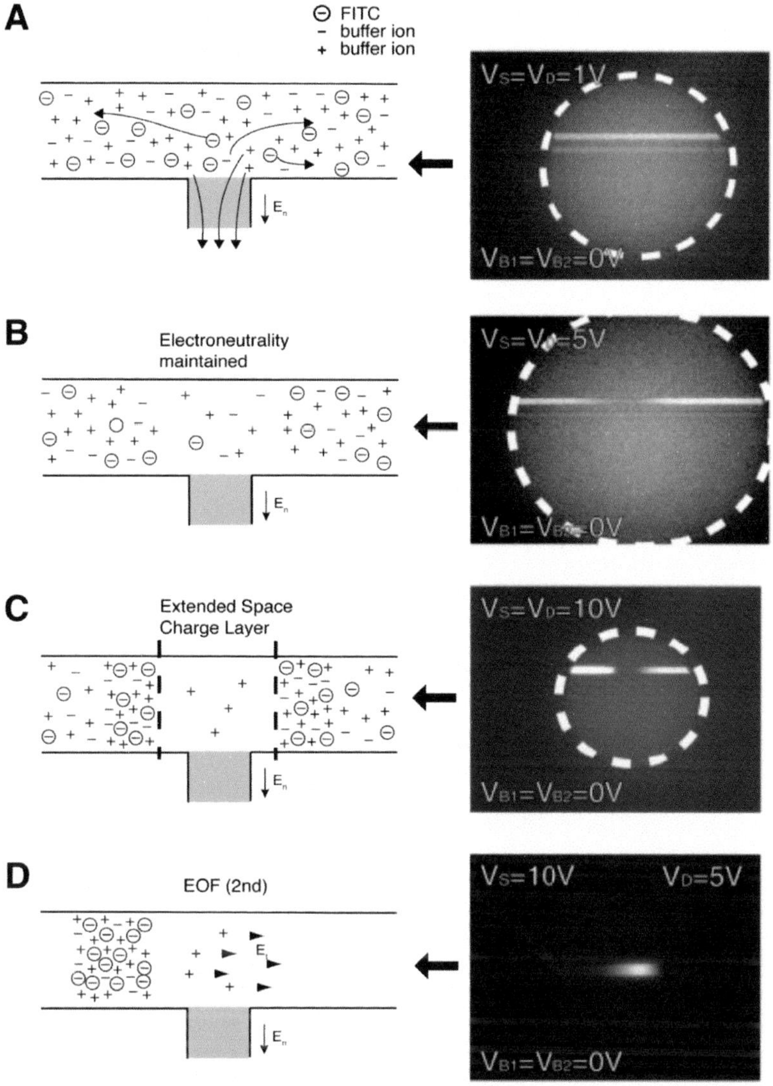

Figure 4.4 Pre-concentration in a hybrid microfluidic/nanofluidic system. (a) When a small trans-nanopore potential, E_n, is applied, ion concentrations in the microchannel are unaffected. (b) At higher values of E_n diffusion limits ion transport across an ion-depletion region. (c) Over time a breakdown in local electroneutrality results, creating a space-charge region above the nanochannels. (d) Application of a second potential along the microchannel, E_t, induces electroosmotic flow of the 2nd kind, rapidly accumulating counterions against the space-charge region. Adapted from Wang, Y.C., A.L. Stevens, and J. Han, *Million-fold preconcentration of proteins and peptides by nanofluidic filter.* Anal Chem, 2005, 77, 4293 and reproduced with permission.

4.4.1 Theory

When flow is electrically driven, the total flux consists of three terms: a diffusive component resulting from the concentration gradient, an electrophoretic component which stems from the electrical potential gradient, and a convective component which originates from the fluid flow. The total flux of the *i-th* species is given by,

$$\Gamma_i = -D_i \nabla c_i - \Omega_i z_i F c_i \nabla \phi + c_i \mathbf{u} \tag{4.8}$$

where F is the Faraday's constant, z_i is the valence, D_i is the diffusion coefficient, Ω_i is the ionic mobility, Γ_i is the flux, c_i is the concentration of the *i-th* species, $\boldsymbol{u}$ is the velocity vector of bulk flow, and φ is the electrical potential. The three terms on the right-hand side of Equation 4.8 define the fluxes due to diffusion, electrical migration, and convection, respectively. The electrical potential distribution is governed by the Poisson equation,

$$\nabla \cdot (\varepsilon_r \nabla \phi) = -\frac{\rho_e}{\varepsilon_0} \tag{4.9}$$

Here ε_0 is the permittivity of free space, ε_r is the relative permittivity, and ρ_e is the net charge density of ions. The net charge density is defined by,

$$\rho_e = F \sum_{i=1}^{m} z_i c_i \tag{4.10}$$

where m is the total number of species involved in the system. The material-balance, or Nernst-Planck (NP), equation describes the mass transfer of each dissolved species and is given by,

$$\frac{\partial c_i}{\partial t} = -\nabla \cdot \Gamma_i \tag{4.11}$$

In micrometer and nanometer channels, the Reynolds number is very low, so the convective terms in the Navier-Stokes equations can be neglected, and the Stokes (Equation 4.12) and continuity (Equation 4.13) equations, which describe the movement of the fluid through the channels, can be solved to obtain the velocity distribution, i.e.

$$-\nabla p + \mu \nabla^2 \mathbf{u} - \rho_e \nabla \phi = 0 \tag{4.12}$$

$$\nabla \cdot \mathbf{u} = 0 \tag{4.13}$$

In the above equations, p is the hydrostatic pressure, and μ is the fluid viscosity. The third term in Equation 4.12 is the body force acting on the fluid due to the net charge density and the electric field. The coupled Poisson and the Nernst-Planck equations (P-NP) and the Stokes equations can be solved using the finite cloud method, and the resulting solutions contain the electric potential, ionic concentration, velocity and pressure profiles.

To understand the physical phenomena in hybrid micro-nanofluidic systems[3,4,6,43] we have considered a canonical problem (Figure 4.5) consisting of a single nanochannel connected to two microchannels.[62] The top microfluidic channel (connected to electrodes 1 and 2) is designated the source channel, and the bottom channel (connected to electrodes 3 and 4) is the receiving channel. The dimensions of the two microchannels are identical, each having a length of 100 μm and a width of 1 μm, while the nanochannel is 1 μm long and 10 nm in width.

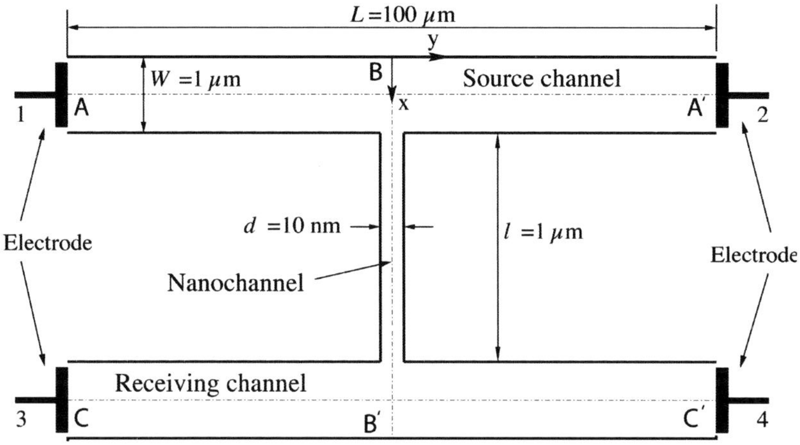

Figure 4.5 Canonical hybrid micro-nanofluidic system not drawn to scale. Note the position of the origin of the coordinate system and the placement of the x- and y- axes. Adapted from Jin, X., *et al.* *Induced electrokinetic transport in micro-nanofluidic interconnect devices.* Langmuir, 2007, **23**, 13209 and reproduced with permission.

The system is designed to control mass transfer between source and receiving channels through the nanochannel. Typical operation of the hybrid micro-nanofluidic system includes 3 stages: rest, injection, and recovery (potential programs for the 3 stages are summarized in Table 4.1). Initially, the system is in the rest stage, then a 25 ms duration injection is initiated, during which ions are transferred from one microchannel to the other. During the injection stage, the source channel is either positively biased relative to the receiving channel or vice versa. Defining the potential bias as $\Delta V = V_{receiving} - V_{source}$, we consider two bias conditions. When $\Delta V > 0$, the electric field is directed from the receiving to the source channel (positively biased case, Table 4.1). When $\Delta V < 0$, the opposite polarity is obtained. After the injection stage, the applied potentials are reverted back to the rest stage, and the system evolves back to the steady-state.

4.4.2 Ion Accumulation and Depletion

Solving the steady-state P-NP and Stokes equations yields the concentration and electrical potential profiles before injection begins. Two-dimensional simulations are justified by assuming that the depth of the micro and nanochannels is much larger than the width, and the channels are filled with 1mM aqueous KH_2PO_4. The surface charge densities of the nanochannels and microchannels are set at +0.35mC/m^2 and 0.05mC/m^2, respectively. Initially the receiving channel is floating, so the electric field in the receiving channel (Figure 4.6(a) top) is ~0. Consequently, there is no ionic migration or EOF (Figure 4.6(a) middle), and thus no current in the receiving channel (Figure 4.6(a) bottom). Figure 4.6(b), shows that the nanochannel anion concentrations are higher than the bulk, with cations being depleted. The Debye length at [KH_2PO_4] = 1 mM is ~10nm ($\kappa^{-1} = 0.308c^{-1/2}$

= 9.74 nm).[62] The electrical double layers in the nanochannel overlap, as illustrated in Figure 4.6(b), leading to an increase in anions and a decrease in cations in the nanochannel, since the surface of the nanochannel is positive.

Potential(V)	Positively biased				Negatively biased			
	1	2	3	4	1	2	3	4
rest	40	0	float	float	40	0	float	float
injection	0	0	40	40	40	40	0	0
recovery	40	0	float	float	40	0	float	float

Table 4.1 Applied voltages during rest, injection and recovery stages.

Transient simulations were performed for two types of injections depending on the applied bias, as shown in Table 4.1. In the positively biased case, i.e. $\Delta V > 0$, Figure 4.7(a) and 4.7(b) show three effects: both cations and anions initially deplete at the source-nanochannel junction region and the depletion region spreads – reaching 40 μm at t = 25 ms, both cations and anions accumulate in the receiving channel with a similar accumulation region width, and a net space charge density is developed in the source microchannel (Figure 4.7(a) bottom), but the receiving channel is electrically neutral (Figure 4.7(b) bottom).

With positive potentials at both ends of the receiving channel, cations tend to move from the receiving channel to the source channel, and conversely for the anions. However, the net cation flux from the ends of the receiving microchannel to the receiving-nanochannel junction is larger than the cation flux from the junction through the nanochannel, because cations are repelled by the positively charged nanochannel. The difference between the fluxes causes the accumulation of cations at the receiving-nanochannel junction. The anion flux from the nanochannel into the receiving channel is greater than that from the receiving channel to the reservoir, because some anions from the nanochannel remain in the junction region to balance the accumulated cations, thus causing an accumulation of anions in the receiving channel. On the other hand, the cation flux from the nanochannel at the source-nanochannel junction is not sufficient to balance the cation flux from the source channel to the reservoirs, due to limited cation transport through the nanochannel. This causes a depletion of cations at the source-nanochannel junction region. Meanwhile, the anion flux from the reservoirs to the source-nanochannel junction is less than the anion flux into the nanochannel, because anions are attracted by the positively charged nanochannel. This difference results in a depletion of anions in the top junction. Thus, both anions and cations are depleted at the source-nanochannel junction, while both anions and cations accumulate at the receiving-nanochannel junction. The Gouy-Chapman model[63] dictates a double layer thickness that varies inversely with electrolyte concentration. Thus, because the ion concentration is extremely low (depleted) at the source-nanochannel junction, the electrical double layer extends a significant distance into the source channel, creating an induced space charge in the source. If the electric field in the nanochannel is reversed, i.e. $\Delta V < 0$, depletion occurs in the receiving channel and accumulation occurs in the source. The ion-enrichment and ion-depletion phenomena are the critical underlying basis for the observed behaviour in nanofluidic-microfluidic composite architectures.[3]

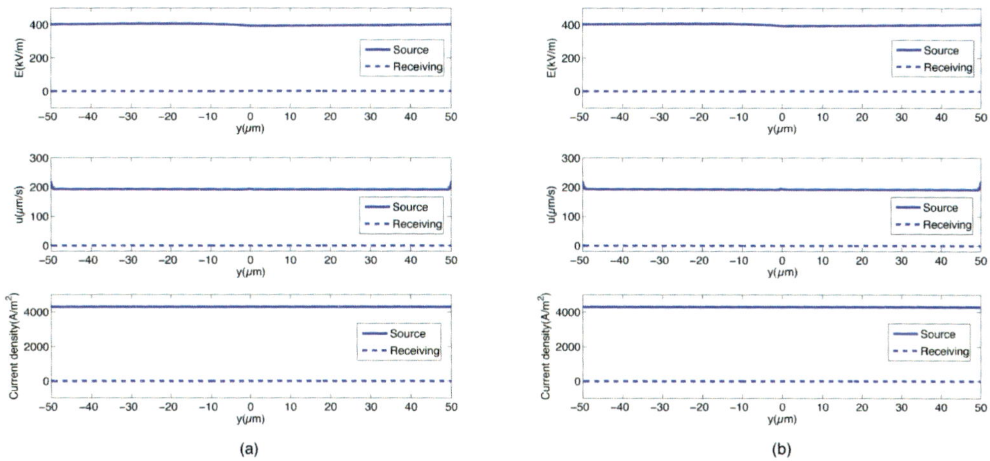

Figure 4.6 (a) Electric field, velocity, and current density in the source and receiving channel. Top: electric field along the central lines AA′ and CC′ (*cf.* Fig. 5). Middle: fluid velocity along the central lines AA′ and CC′. Bottom: current density along the central lines AA′ and CC′. (b) Ionic concentrations and potential across the nanochannel at x = 1.5μm. Top: cation and anion concentrations. Bottom: electric potential.

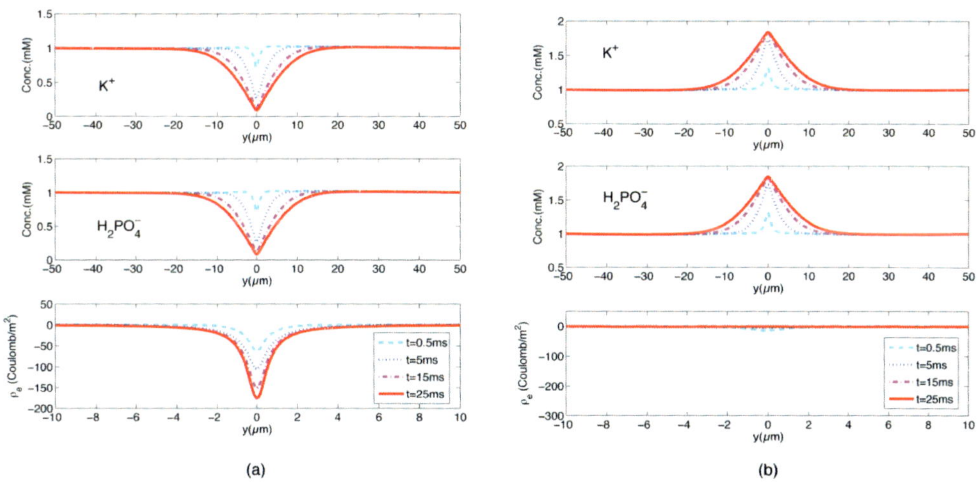

Figure 4.7 Simulation of the injection stage with ΔV = 40V potential bias: (a) Ionic concentrations along the central line AA′ of the source channel. Top: cation concentration. Middle: anion concentration. Bottom: the space charge density. (b) Ionic concentrations along the central line CC′ of the receiving channel. Top: cation concentration. Middle: anion concentration. Bottom: the space charge density.

4.4.3 Ionic Currents

Figure 4.8 shows the transient variation of the ionic current during the rest, injection and recovery stages when the system is positively biased (Table 4.1). The current signature shows that when injection is initiated, the current drops to less than 10% of the rest current within 25 ms, then during the recovery stage, the current initially increases rapidly, reaching a plateau, before a gradual increase to the steady-state value. During injection at positive bias, ions deplete at the source-nano junction and accumulate at the receiving-nano junction. Depletion increases electrical resistance in the source-nano junction, while accumulation decreases resistance in the receiving-nano junction. However, depletion has a larger effect on resistance than accumulation, because resistance is inversely proportional to the ionic concentration. Thus, the overall resistance in the system increases, causing the ionic current to decrease.

At $t = 25$ ms, when the electrode potentials are switched back to the rest state, two key features are observed: (*cf* Figure 4.9), the depletion region broadens and moves out of the source channel due to fluid flow in the source channel. The ionic current increases rapidly as the source channel current is re-established, and then reaches a plateau, where the current is limited by the depletion region conductivity. When the leading edge of the depletion region reaches the end of the source channel, the ionic current increases again as the depletion region exits the channel. If the direction of injection is reversed by applying a negative bias (Table 4.1), the ionic current signature in the source channel again shows biphasic behaviour, but with an initial high current relaxing to a smaller steady-state value - behaviour which can be explained by the converse of the behaviour at positive bias.

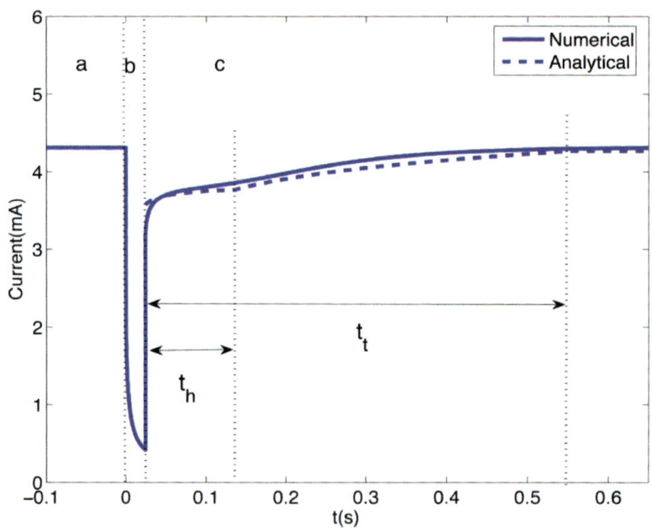

Figure 4.8 Ionic current in the source channel *vs.* time when the system is positively biased. (a) current during the rest stage, (b) current during the injection stage, and (c) current during the recovery stage. The analytical solution is calculated from equation 4.16 in reference [62]. Adapted from Jin, X., *et al. Induced electrokinetic transport in micro-nanofluidic interconnect devices.* Langmuir, 2007, **23**, 13209 and reproduced with permission.

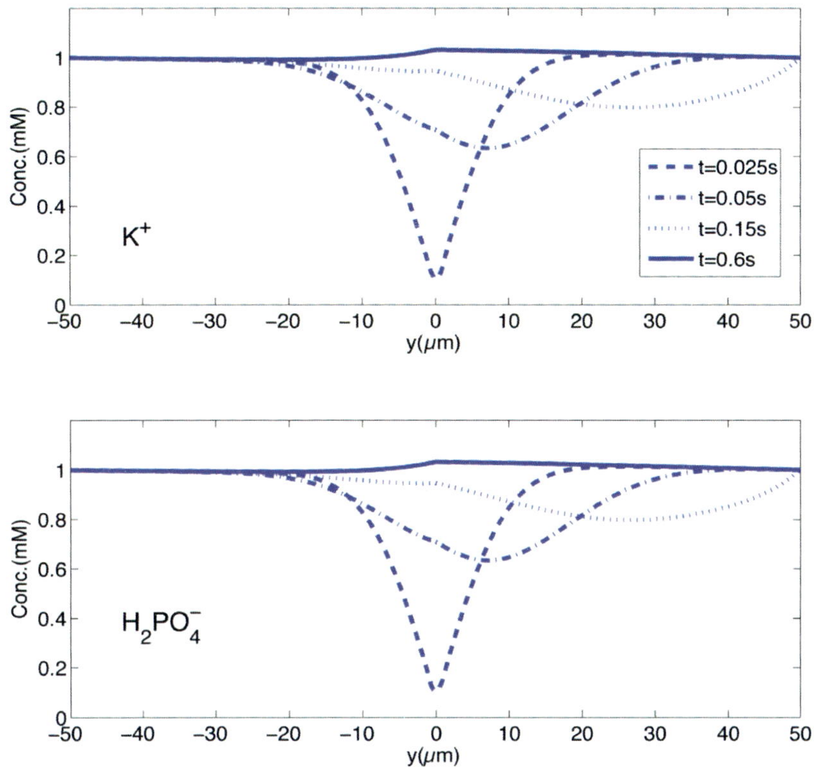

Figure 4.9 Ion concentration along the central line of the source channel during the recovery stage. Note that the recovery stage starts from $t = 25$ ms. (Top) Cation concentration. (Bottom) Anion concentration.

4.4.4 Induced Flow

EOF and induced pressure driven flow are coupled during the injection stage resulting in induced space charges in the junction region that interact with an external electric field in the microchannel to determine the fluid velocity in the microchannel. During the injection stage when the system is positively biased, three effects are observed: EOF in the nanochannel generates induced pressure-driven flow in the microchannel; fluid flow at the center of the microchannel is dominated by the induced pressure-driven flow; and fluid circulates in the junction region. At $\kappa a \sim 1$ when the electrical double layers overlap and the counter ion concentration is sufficiently large, EOF in the positively charged nanochannel dominates EOF caused by the negatively charged microchannel, causing fluid flow from the source to the receiving channel. The direction of EOF in the nanochannel and the microchannels, and the direction of the induced pressure driven flow in the microchannels, are shown in Figure 4.10(a). The fluid flow in the microchannel is determined by both the vector sum of EOF and the induced pressure driven flow. For example, in the receiving channel the pressure pushes the fluid from the junction region to the reservoir, while the direction of the EOF is from the reservoir to the junction. The net result is that the fluid flow at the center of the microchannel is in the same direction as the pressure-driven flow, but in the opposite direction near the microchannel surface, cf. Figure 4.10(b). The velocity profile is similar to that typically observed when pressure driven flow and EOF combine.[64] Figure 4.11 shows the streamlines near the micro-nano junctions, with vortices near the entrance of the nanochannel resulting from the opposing directions of EOF and induced pressure driven flow in the microchannel. The ratio of the pressure gradient to the electrical body force is ~1 in the junction regions at t = 15 ms during injection, giving rise to a recirculating eddy.[65] Such circulation of fluid can enhance mixing of analytes, explaining the rapid mixing achieved in experiments with micro-nano devices.[6]

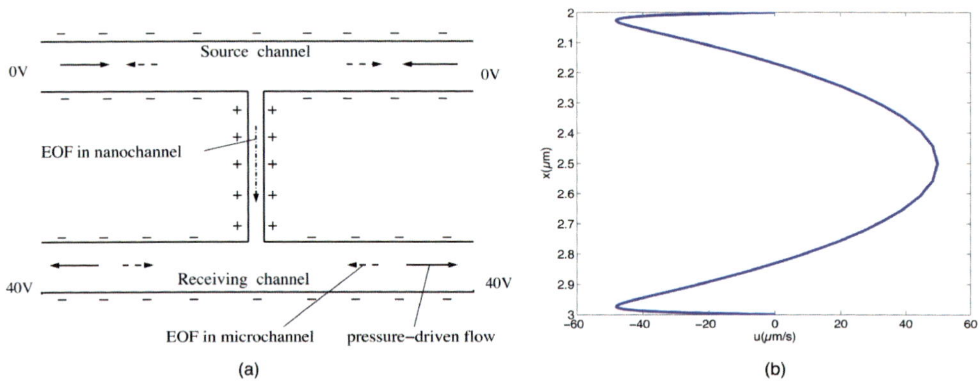

Figure 4.10 Electroosmotic flow (EOF) during the injection stage when the receiving channel is positively biased relative to the source channel. (a) Schematic diagram indicating the direction of electroosmotic flow and induced pressure-driven flow in the system. (b) Velocity profile across the receiving channel near the reservoir (at t = 15 ms, y = 45 μm). Adapted from Jin, X., *et al. Induced electrokinetic transport in micro-nanofluidic interconnect devices.* Langmuir, 2007, **23**, 13209 and reproduced with permission.

Although EOF normally results from the electrical double layer at the surface, Dukhin[66] predicted that the space charges induced by an external electric field could also lead to fluid flow and called it electroosmosis of the second kind. This EOF of the second kind has been observed adjacent to ion-permselective particles,[67-69] where the space charges induced by the applied field on the surface of the particles spread over a larger region than the primary electrical double layer, and give rise to highly chaotic flow patterns in the extra-particle void space. High average sample flow speed, ~ 1 mm/s measured in a protein pre-concentration device where the expected electroosmotic velocity due to surface charges alone is 10-100 μm/s, has been attributed to EOF of the second kind resulting from induced space charges.[57] Space charges spread over the micro-nano junction region and are capable of inducing EOF in the microchannel as observed in the calculations. During the injection stage when the system is positively biased, the electric field in the nanochannel induces a space charge at the source-nano junction region, as shown in Figure 4.7(a) (bottom). After the system is switched back to the recovery stage, the net negative space charge redistributes, the charge distribution along the center line of the source microchannel being shown in Figure 4.12. After injection, the source-nanochannel junction is depleted of both cations and anions, but satisfying electroneutrality at steady state requires more anions than cations. At the start of recovery, anions in the source channel enter the nanochannel from one side and move toward electrode 1 from the other, such that the anion concentration on the left side of the source-nano junction region (t = 26 ms, -20 μm < y < 0 μm) is lower than that of the cations, causing the observed asymmetric charge distribution. As recovery progresses the asymmetry dissipates.

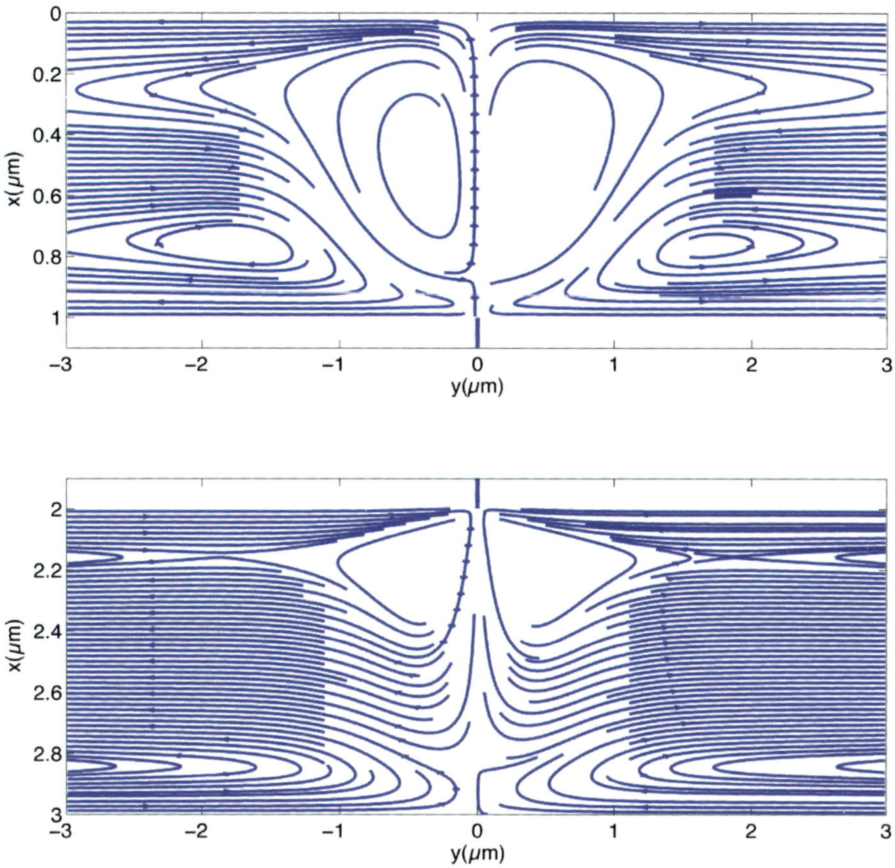

Figure 4.11 Streamlines (at t = 15ms) near the micro-nano junction regions showing the vortices developed due to opposing electroosmotic and pressure driven flows in the microchannel. Top: near the junction of the source channel and the nanochannel. Bottom: near the junction of the receiving channel and the nanochannel. Adapted from Jin, X., *et al. Induced electrokinetic transport in micro-nanofluidic interconnect devices.* Langmuir, 2007, **23**, 13209 and reproduced with permission.

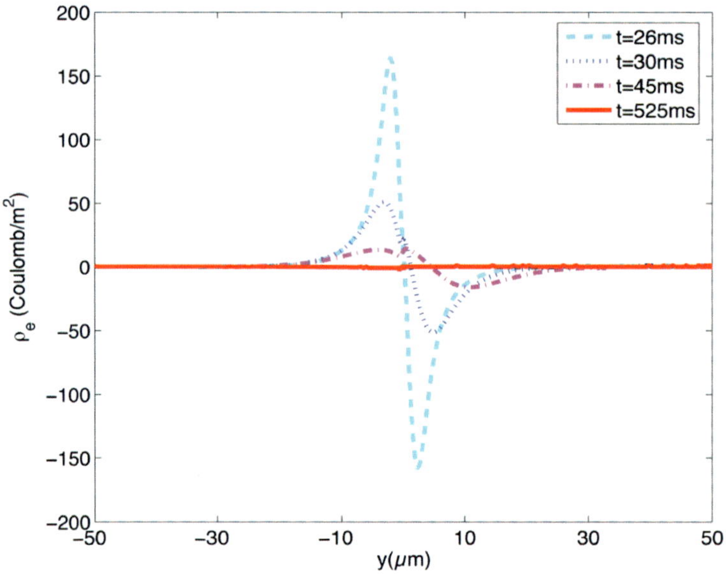

Figure 4.12 Net charge density along the center line of the source channel during the recovery stage. Adapted from Jin, X., *et al. Induced electrokinetic transport in micro-nanofluidic interconnect devices.* Langmuir, 2007, **23**, 13209 and reproduced with permission.

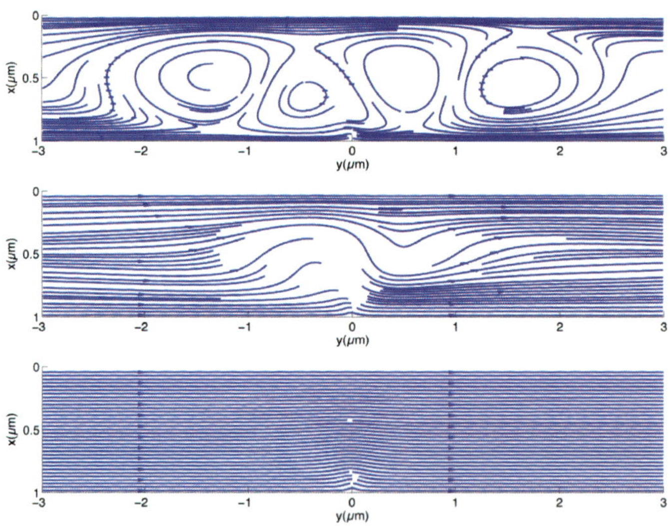

Figure 4.13 The streamlines near the source-nanochannel junction region during the recovery stage. Top: at t = 26ms. Middle: at t = 30ms. Bottom: at t = 525ms. Adapted from Jin, X., *et al. Induced electrokinetic transport in micro-nanofluidic interconnect devices.* Langmuir, 2007, **23**, 13209 and reproduced with permission.

The induced space charges in the microchannel cause EOF of the second kind in the source microchannel in addition to the EOF caused by the microchannel surface charge. Figure 4.13 shows the streamlines of flow in the source channel caused by induced charges near the source-nanochannel junction. At t = 26 ms, vortices are generated as the charge redistributes. The space charges induce a large pressure gradient in the junction region (cf. Figure 4.14(a)). Near electrode 2, at y = 45 μm, the net flow rate is more than 2.5 times the steady state EOF (cf. Figure 4.14(b)). As recovery progresses, the induced space charge is dispersed, and its effect on the net flow diminishes. Although both the depletion zone and the induced space charge contribute to the flow, the major contribution is from the space charges. As shown in Figure 4.9, the depletion region is still in the microchannel at t = 50 ms, but at t = 45 ms the velocity (Figure 4.14(b)) is almost the same as the steady-state velocity. However, because the magnitude of the induced space charge is proportional to the number of nanochannels and their surface charge densities, using a membrane consisting of 1000 nanochannels extremely large flow rates could be attained.

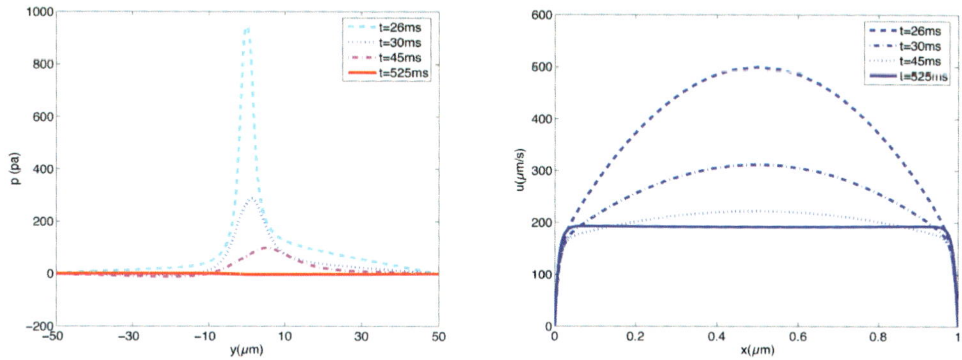

Figure 4.14 Pressure and velocity profiles during the recovery stage. (a): Pressure along the central line of the source channel. (b): Velocity profile across the source channel (at y = 45μm). Adapted from Jin, X., *et al. Induced electrokinetic transport in micro-nanofluidic interconnect devices.* Langmuir, 2007, **23**, 13209 and reproduced with permission.

4.5 CONCLUSIONS

Clearly, nanofluidic systems facilitate the extension of traditional μ-TAS systems both spatially (i.e. nanofluidic interconnects allow 3-D μ-TAS structures) and functionally. While a basic taxonomy of nanofluidic flow phenomena has been determined experimentally, a detailed chemical understanding is more fully achieved by synergistically coupling experiment and modelling. This is especially true for integrated microfluidic/nanofluidic systems where the interplay between microfluidic and nanofluidic elements results in behaviour not observed in either system independently. The case of ion enrichment/depletion in hybrid devices clearly demonstrates these principles.

4.6 ACKNOWLEDGEMENTS

Work described in this chapter carried out in the authors' laboratories was supported by the National Science Foundation through the Science and Technology Center for Advanced

Materials for Water Purification with Systems (CTS-0120978), the Nano-CEMMS Center (DMI-0328162), and by the Department of Energy under grant DE FG02 07ER15851.

References

1. K. Fa, J. J. Tulock, J. V. Sweedler and P. W. Bohn, *J. Am. Chem. Soc.*, 2005, **127**, 13928-13933.
2. D. M. Cannon, Jr., T. C. Kuo, P. W. Bohn and J. V. Sweedler, *Anal. Chem.*, 2003, **75**, 2224-2230.
3. T. C. Kuo, D. M. Cannon, M. A. Shannon, P. W. Bohn and J. V. Sweedler, *Sens. Actuators A*, 2003, **102**, 223-233.
4. J. J. Tulock, M. A. Shannon, P. W. Bohn and J. V. Sweedler, *Anal. Chem.*, 2004, **76**, 6419-6425.
5. M. Davidson, P. Dommersnes, M. Markstrom, J. F. Joanny, M. Karlsson and O. Orwar, *J. Am. Chem. Soc.*, 2005, **127**, 1251-1257.
6. T. C. Kuo, H. K. Kim, D. M. Cannon, M. A. Shannon, J. V. Sweedler and P. W. Bohn, *Angew. Chem., Int. Ed.*, 2004, **43**, 1862-1865.
7. J. O. Tegenfeldt, C. Prinz, H. Cao, R. L. Huang, R. H. Austin, S. Y. Chou, E. C. Cox and J. C. Sturm, *Anal. Bioanal. Chem.*, 2004, **378**, 1678-1692.
8. J. P. Fu, P. Mao and J. Y. Han, *App. Phys. Lett.*, 2005, **87**, 263902.
9. J. P. Fu, R. B. Schoch, A. L. Stevens, S. R. Tannenbaum and J. Y. Han, *Nat. Nanotechnol.*, 2007, **2**, 121-128.
10. J. P. Fu, J. Yoo and J. Y. Han, *Phys. Rev. Lett.*, 2006, **97**, 018103.
11. Z. Siwy, L. Trofin, P. Kohli, L. A. Baker, C. Trautmann and C. R. Martin, *J. Am. Chem. Soc.*, 2005, **127**, 5000-5001.
12. T. Ito, L. Sun, R. R. Henriquez and R. M. Crooks, *Acc. Chem. Res.*, 2004, **37**, 937-945.
13. A. S. Bhown and P. Stroeve, *Ind. Eng. Chem. Res.*, 2007, **46**, 6118-6125.
14. R. Karnik, R. Fan, M. Yue, D. Li, P. Yang and A. Majumdar, *Nano Lett.*, 2005, **5**, 943-948.
15. S. U. Hong, R. Malaisamy and M. L. Bruening, *Langmuir*, 2007, **23**, 1716-1722.
16. S. B. Lee and C. R. Martin, *Anal. Chem.*, 2001, **73**, 768-775.
17. J. M. Iannacone, J. A. Jakubowski, P. W. Bohn and J. V. Sweedler, *Electrophoresis*, 2005, **26**, 4684-4690.
18. R. B. Schoch, L. F. Cheow and J. Han, *Nano Lett.*, 2007, **7**, 3895-3900.
19. B. Y. Kim, C. B. Swearingen, J. A. A. Ho, E. V. Romanova, P. W. Bohn and J. V. Sweedler, *J. Am. Chem. Soc.*, 2007, **129**, 7620-7626.
20. E. N. Gatimu, T. L. King, J. V. Sweedler and P. W. Bohn, *Biomicrofluidics*, 2007, **1**, 021502.
21. M. Nishizawa, V. P. Menon and C. R. Martin, *Science*, 1995, **268**, 700-702.
22. H. Bayley and C. R. Martin, *Chem. Rev.*, 2000, **100**, 2575-2594.
23. L. T. Sexton, L. P. Horne, S. A. Sherrill, G. W. Bishop, L. A. Baker and C. R. Martin, *J. Am. Chem. Soc.*, 2007, **129**, 13144-13152.
24. A. J. Storm, J. H. Chen, H. W. Zandbergen and C. Dekker, *Phys. Rev. E: Stat. Phys., Plasmas, Fluids, Relat. Interdiscip. Top.*, 2005, **71**, 051903.
25. C. L. Rice and R. Whitehead, *J. Phys. Chem.*, 1965, **69**, 4017-4024.
26. K. Y. Chun and P. Stroeve, *Langmuir*, 2002, **18**, 4653-4658.

27. M. Wirtz, S. F. Yu and C. R. Martin, *Analyst*, 2002, **127**, 871-879.
28. J. R. Ku, S. M. Lai, N. Ileri, P. Ramirez, S. Mafe and P. Stroeve, *J. Phys. Chem. C*, 2007, **111**, 2965-2973.
29. R. B. Schoch, A. Bertsch and P. Renaud, *Nano. Lett.*, 2006, **6**, 543-547.
30. K. Y. Chun, S. Mafe, P. Ramirez and P. Stroeve, *Chem. Phys. Lett.*, 2006, **418**, 561-564.
31. N. C. Li, S. F. Yu, C. C. Harrell and C. R. Martin, *Anal. Chem.*, 2004, **76**, 2025-2030.
32. Z. Siwy, E. Heins, C. C. Harrell, P. Kohli and C. R. Martin, *J. Am. Chem. Soc.*, 2004, **126**, 10850-10851.
33. A. L. Garcia, L. K. Ista, D. N. Petsev, M. J. O'Brien, P. Bisong, A. A. Mammoli, S. R. J. Brueck and G. P. Lopez, *Lab Chip*, 2005, **5**, 1271-1276.
34. D. Stein, M. Kruithof and C. Dekker, *Phys. Rev. Lett.*, 2004, **93**, 035901.
35. J. R. Ku and P. Stroeve, *Langmuir*, 2004, **20**, 2030-2032.
36. R. Karnik, K. Castelino, R. Fan, P. Yang and A. Majumdar, *Nano Lett.*, 2005, **5**, 1638-1642.
37. A. Plecis, R. B. Schoch and P. Renaud, *Nano Lett.*, 2005, **5**, 1147-1155.
38. R. B. Schoch and P. Renaud, *Appl. Phys. Lett.*, 2005, **86**, 253111.
39. R. B. Schoch, H. van Lintel and P. Renaud, *Phys. Fluids*, 2005, **17**, 100604.
40. Z. Siwy, I. D. Kosinska, A. Fulinski and C. R. Martin, *Phys. Rev. Lett.*, 2005, **94**, 48102.
41. P. J. Kemery, J. K. Steehler and P. W. Bohn, *Langmuir*, 1998, **14**, 2884-2889.
42. T. C. Kuo, L. A. Sloan, J. V. Sweedler and P. W. Bohn, *Langmuir*, 2001, **17**, 6298-6303.
43. T. C. Kuo, D. M. Cannon, Jr., Y. Chen, J. J. Tulock, M. A. Shannon, J. V. Sweedler and P. W. Bohn, *Anal. Chem.*, 2003, **75**, 1861-1867.
44. I. H. Chang, J. J. Tulock, J. Liu, W. S. Kim, D. M. Cannon, Jr., Y. Lu, P. W. Bohn, J. V. Sweedler and D. M. Cropek, *Environ. Sci. Technol.*, 2005, **39**, 3756-3761.
45. J. S. Kirk, J. V. Sweedler and P. W. Bohn, *Anal. Chem.*, 2006, **78**, 2335-2341.
46. S. D. Noblitt, J. R. Kraly, J. M. VanBuren, S. V. Hering, J. L. Collett and C. S. Henry, *Anal. Chem.*, 2007, **79**, 6249-6254.
47. B. Bauer, M. Davidson and O. Orwar, *Langmuir*, 2006, **22**, 9329-9332.
48. J. Hurtig, B. Gustafsson, M. Tokarz and O. Orwar, *Anal. Chem.*, 2006, **78**, 5281-5288.
49. A. Karlsson, M. Karlsson, R. Karlsson, K. Sott, A. Lundqvist, M. Tokarz and O. Orwar, *Anal. Chem.*, 2003, **75**, 2529-2537.
50. R. Karlsson, A. Karlsson and O. Orwar, *J. Am. Chem. Soc.*, 2003, **125**, 8442-8443.
51. R. Karlsson, A. Karlsson and O. Orwar, *J. Phys. Chem. B*, 2003, **107**, 11201-11207.
52. J. Khandurina, S. C. Jacobson, L. C. Waters, R. S. Foote and J. M. Ramsey, *Anal. Chem.*, 1999, **71**, 1815-1819.
53. R. S. Foote, J. Khandurina, S. C. Jacobson and J. M. Ramsey, *Anal. Chem.*, 2005, **77**, 57-63.
54. R. Dhopeshwarkar, S. A. Li and R. M. Crooks, *Lab Chip*, 2005, **5**, 1148-1154.
55. Y. Zhang and A. T. Timperman, *Analyst*, 2003, **128**, 537-542.
56. Z. Long, D. Liu, N. Ye, J. Qin and B. Lin, *Electrophoresis*, 2006, **27**, 4927-4934.
57. Y. C. Wang, A. L. Stevens and J. Han, *Anal Chem*, 2005, **77**, 4293-4299.
58. S. J. Kim, Y. C. Wang, J. H. Lee, H. Jang and J. Han, *Phys. Rev. Lett.*, 2007, **99**, 044501.
59. J. H. Lee, S. Chung, S. J. Kim and J. Y. Han, *Anal. Chem.*, 2007, **79**, 6868-6873.

60. P. Jain, L. Sun, J. H. Dai, G. L. Baker and M. L. Bruening, *Biomacromolecules*, 2007, **8**, 3102-3107.
61. L. Sun, J. H. Dai, G. L. Baker and M. L. Bruening, *Chem. Mater.*, 2006, **18**, 4033-4039.
62. X. Jin, S. Joseph, E. Gatimu, P.W. Bohn and N. Aluru, *Langmuir* 2007, **23**, 13209-13222.
63. A. V. Delgado and F. J. Arroyo, in *Interfacial Electrokinetics and Electrophoresis*, ed. A. V. Delgado, CRC Press, 1 st edn., 2002, pp. 1-54.
64. G. E. Karniadakis, A. Beskok and N. R. Aluru, *Microflows and Nanoflows: Fundamentals and Simulation*, Springer, 2005.
65. S. Y. Park, C. J. Russo, D. Branton and H. A. Stone, *J. Colloid Interface Sci.,* 2006, **297,** 832-839.
66. S. S. Dukhin, *Adv. Colloid Interface Sci.,* 1991, **35,** 173-196.
67. F. C. Leinweber and U. Tallarek, *Langmuir,* 2004, **20,** 11637-11648.
68. F. C. Leinweber and U. Tallarek, *J. Phys. Chem. B,* 2005, **109,** 21481-21485.
69. N. A. Mishchuk and S. S. Dukhin, in *Interfacial Electrokinetics and Electrophoresis,* ed. A. V. Delgado, CRC Press, 1st edn., 2002, p. 241-276.

CHAPTER 5

Fabrication of Silica Nanofluidic Tubing for Single Molecule Detection

MIAO WANG[1] & JUN KAMEOKA[1]

[1] Department of Electrical and Computer Engineering, Texas A&M University, College Station, Texas, TX 77843-3128, U.S.A

5.1 INTRODUCTION

Nanoscale fluidic systems have gained increasing attention in recent years because their representative small dimensions are comparable to the size of biomolecules such as proteins and DNA. Accordingly, they are expected to provide new ways to manipulate or detect such molecules with unrivalled precision. For instance, they have been used in DNA analysis,[1] single molecule detection[2,3] and fluorescence correlation spectroscopy[2].

Fluorescence spectroscopy has been used as a powerful tool for detecting molecules in a variety of chemical and biological systems for many years.[4] Fluorescence signals from specific fluorophores can be recorded and spectral, intensity, lifetime or polarisation information gained from this can be used for molecular detection or identification. For the detection of single molecules in fluidic media, photon burst signals from single molecules traversing a sub-femtoliter optical probe volume can be used with auto-correlation or cross-correlation analysis for both molecule sizing and identification. Indeed, by fitting such correlation curves to theoretical models, much information relating to chemical kinetics and molecular mobility may be obtained. This so-called fluorescence correlation spectroscopy (FCS) is the most common method for detecting single molecules in liquid phase environments.[5-28] Nevertheless, conventional FCS methods normally generate information representative of all molecules passing through a given detection volume, rather than providing information about individual molecules. In recent years, FCS-based techniques have been used in conjunction with microfluidic systems.[29,30] Laser light is focused in the centre of a microchannel, and the fluorescence signals originating from individual molecules passing through this probe volume are detected by an avalanche photodiode. Compared to conventional FCS, the incorporation of a microfluidic system for sample manipulation has several advantages. First, the statistical accuracy of single-molecule characterization is improved because all (or the vast majority of) molecules are counted and contribute to the statistical analysis. In addition, detection throughput can be controlled in a facile manner. The number of molecules detected per unit time can simply be increased by increasing the electrical potential or flow pumping speed. Significantly, this control can be used to drastically reduce the assay time. Moreover, because of the small detection volume, the signal-to-noise (S/N) ratio is improved.[31-36] To further improve the S/N ratio, nanochannels can be utilized for single molecule detection since they act to decrease the size of detection probe volume. A reduced detection volume will significantly improve the S/N ratio and decrease the probability of multiple molecule occupancy of the detection volume. For instance, the S/N ratio obtainable in 800 nm square-shaped channels is 5 times better than 20 μm square-shaped channels. However, the smaller nanochannel is expected to provide rather different molecular transport mechanisms.[37-40] According to these investigations, the electrokinetic molecular mobility in a microchannel is dominated by the electroosmotic flow that is independent of molecular charge. Conversely, studies by Garcia,[37] Pennathur[39] and Karniadakis[40] have shown that molecule mobility in nanofluidic channels becomes a strong function of molecular charge. Thus, the difference in mobility between molecules of similar structure, such as phosphorylated and normal molecules (containing one charge difference), should be more apparent.

In this chapter, the fabrication of nanofluidic conduits including lithography-based top-down techniques and bottom-up methods will be discussed. Subsequently, the advantages of nanofluidic conduits for single molecule detection will be demonstrated with reference to reported experimental studies.

5.2 FABRICATION OF SILICA NANOFLUIDIC TUBES

5.2.1 Concepts

As the name suggest, nanofluidic tubing is characterised by its extremely small (nanometer scale) cross-sectional dimension, which as noted is a favourable feature for use in single molecule detection applications. By spatially confining a limited amount of target molecules within an ultra-small detection volume, nanofluidic tubing serves as a reliable and powerful analysis platform. The practical application of such a technology is highly dependent on the development of innovative fabrication techniques that can be used to create bore diameters on the nanometer scale. Although a diversity microfabrication techniques have been developed and matured over the last 50 years,[41] nanofabrication method development is a more nascent research field and is currently being explored by many researchers. The vast majority of these researches are focused on how the resolution limitations of conventional micromachining methods may be overcome, whilst reducing fabrication cost and increasing device throughput.

Nanofluidic tubing fabrication techniques can be broadly divided into two categories: top-down and bottom-up methods.[42] Lithographic processes including both light and electron beam methods define the standard top-down approach. These are widely used for device manufacture in the semiconductor industries. Within this general approach, a nanofluidic channel may be defined using lithography, followed by reactive ion etching and then sealing or bonding of the structured substrate to a cover.[43] The 'nano' aspect of the channel or conduit (i.e. the cross-sectional dimensions) is lithographically defined during the resist patterning process. Unfortunately, this approach normally demands the use of high resolution lithography techniques, such as X-ray lithography,[44] e-beam lithography,[45,46] and proton beam lithography.[47] All such techniques can overcome the diffraction limit of light and thus improve feature resolution. However, the cost of ownership and system maintenance is extremely high. E-beam and proton beam lithography allows flexible pattern generation within a photoresist by using a 'mask-less' direct beam writing process. However the serial nature of this process also causes low throughout and thus makes it unsuitable for mass manufacturing. Focused-ion-beam milling[48,49] is another kind of direct writing process which rasters over and directly patterns a material. Needless to say, the process is not high-throughput in nature and is costly.

Other lithographic techniques such as interferometric lithography[50] and nanoimprint lithography[51,52] have been assessed as tools for fabricating nanofluidic channels. Interferometric lithography uses a standing wave pattern generated by multiple coherent optical beams to expose a photoresist layer. One of the primary challenges when using this approach is realizing pattern flexibility, since two-beam interferometric lithography will only produce periodic lines and spaces. Multiple beams and multiple exposures are currently being explored to create more complex structures.[53,54] Nanoimprint lithography is a simple process which transfers a pattern from a template to an imprint resist by mechanical deformation. Its utility depends on the structure of template, which is usually fabricated using ion beam lithography or electron beam lithography.

In addition to lithographic processing, an etching step is required to transfer the lithographically patterned nano-scale tubing on the resist into a substrate such as fused silica or silicon. In this step, a tubing depth in the nanometer range must be established. For instance, reactive ion etching can be controlled to produce a nanofluidic channel with a sub-100 nm depth on the surface of borosilicate[55] and silicon.[56] Wet anisotropic etching has also been used to develop nanofluidic channels as shallow as 50nm on <110> silicon wafer using native oxide as the mask and Olin OPD 4262 positive resist developer as the etchant.[43]

The last step in top-down fabrication of a nanofluidic channel is sealing or bonding. This process is used to enclose the etched trenches. There are a variety of ways to achieve a seal such as anodic bonding, fusion bonding, polymer bonding and eutectic bonding.[42] The selection of a sealing or bonding method depends sensitively on the substrate material being used. In addition, some sealing methods can narrow down the channel size. For example, Austin and co-workers reported the sealing of a nanofluidic channel array generated by nanoimprint lithography by depositing SiO_2 over the trenches at a wide distribution of angles to create a capping layer. The local shadowing effects inherent in the deposition process reduced the cross section to approximately 10 nm.[51]

Self-sealed nanofluidic tubes have been fabricated by many researchers. For instance, sacrificial layer methods have been commonly used.[46,57] This approach involves patterning sacrificial materials on the surface of a substrate using high resolution pattering techniques such as e-beam or nanoimprint lithography. Subsequently a capping layer is deposited to cover the patterned sacrificial structure. In the final step, the sacrificial layer is removed using chemical solvents or thermal decomposition. By using this approach, nanofluidic tubes with uniform height can be produced. However the removal of sacrificial material normally occurs over an extended time period, with conduit dimensions being sensitive to the pattering techniques used.

In general, top-down nanochannel fabrication methods are normally costly and generate rectangular or triangular shape cross-sectional profiles. Additionally, the success of device fabrication relies completely on the lithography process. Interestingly, reports relating to the bottom-up fabrication of nanofluidic tubes are rarer, due to the difficulty in controlling the self-assembly process. In this process, atoms and molecules are initiated to arrange themselves into more complex nanostructures. Bottom-up approaches to nanochannel formation are however attractive due to their low cost and ability to create nanostructures with extremely small dimensions. Electrospinning, which is a simple and viable technique to fabricate continuous micro/nanofibers directly from a polymer/polymer-blended solution, is a classical self-assembly process. However opinions are split about whether electrospinning is a top-down or bottom-up fabrication method.[58,59] Details concerning this technique are discussed in the proceeding paragraphs.

5.2.2 Electrospinning

5.2.2.1 Basics of Electrospinning

In 1934, Anton Formhals invented a spinning technology that produced synthetic fibers with the aid of an electric field.[60] This invention did not gain widespread attention until the 1990s when researchers demonstrated its capacity to produce nanoscale polymer fibers.[61]

A typical electrospinning process involves the extrusion of a polymer solution from a needle or spinneret by applying a high voltage. A so-called "Taylor cone" is formed at the tip of the spinneret since the induced charges in the polymer solution are attracted by the electrostatic force. A grounded collector or a counter electrode is placed several centimeters away from the spinneret or source. When the applied voltage exceeds a threshold value, a polymer jet is extruded because the electrostatic force overcomes the surface tension of polymer solution, and the polymer erupts from the Taylor cone and travels towards the grounded collector, which is a lower potential zone, under an electrostatic stretching force. In this procedure, the liquid jet experiences a bending and whipping motion due to charge repulsion between molecules incorporated in the liquid jet. The repulsion of electric charges in the jet, which continuously stretches and minimizes the diameter of the jet, together with the rapidly evaporating solvent, leads to size-shrinking and solidification of micro/nanofibers on the grounded collector. A piece of aluminium foil or a silicon wafer may be attached to the collector for collecting nanofibers. A schematic diagram of a basic electrospinning process is shown in Figure 5.1.

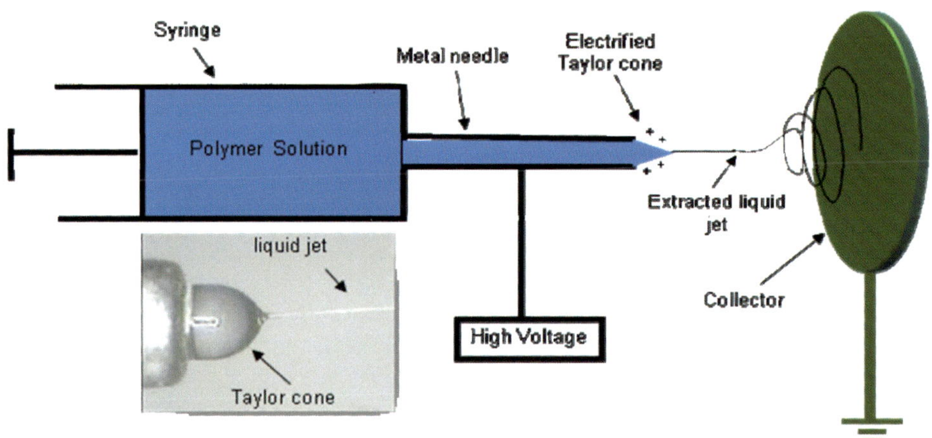

Figure 5.1 Schematic diagram of a basic electrospinning setup. The inset shows an image of an electrified Taylor cone and an extracted liquid jet

Nanofibers with a solid interior and a smooth surface can be fabricated by a normal electrospinning process. An appropriate polymer solution is a prerequisite for this technique. Electrospinning is able to produce long, thin and continuous fibers from various materials including polymers, ceramics and composites. A broad range of polymers have been directly electrospun into a nanofiber format, including polyvinyl pyrrolidone (PVP),[61] poly(vinylidenefluoride) (PVDF),[62,63] poly(caprolactone),[64,65] poly(glycolide) (PGA),[66] poly(L-lactide) (PLA),[66] polystyrene (PS),[67] polyethylene oxide (PEO),[65] polyvinyl alcohol (PVA), etc.[65] Because of the simplicity and low-cost of electrospinning, much effort has been focused on expanding the scope of 'electrospinable' materials and associated applications.

By blending sol-gel precursors with polymers such as PVP, PVA and PEO, metal oxide nanofibers, ceramic nanofibers and even non-metal-oxide ceramic fibers can be produced effectively.[61] Figure 5.2 shows examples of ZrO_2 and SiO_2 nanofibers fabricated by electrospinning. Carbon nanotubes, nanoparticles, drugs, dyes, enzymes and DNA have also been incorporated into nanofibers successfully. The large surface area-to-volume ratio is especially favourable for many applications such as biomedical scaffolds in tissue engineering, chemical catalysts, or sensors.[61]

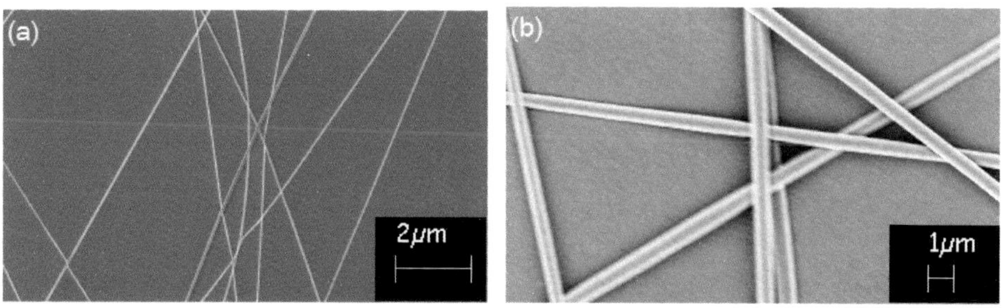

Figure 5.2 SEM images of (a) ZrO_2 nanofibers (b) SiO_2 nanofibers fabricated by electrospinning

Various factors can influence the electrospinning process. The primary processing parameters involve: (1) *The properties of polymer solution* (e.g. viscosity, electrical conductivity and surface tension). The viscosity can be controlled by varying the polymer type and concentration. Different solvents and additives can be used to control the electrical conductivity and surface tension of the solution. (2) *The applied voltage*, which affects the electric field between the nozzle and the collector, can be adjusted using an external power supply. The applied voltage controls the extrusion force of the liquid jet, which must be balanced with the surface tension and viscous forces of the polymer solution. (3) *The flow rate*, which defines the rate that the electrospun polymer solution is supplied. This can be adjusted by a mechanical pump connected to a syringe, and controls the volume of solution emerging from the nozzle per unit time. (4) *The deposition distance*, which is the distance between the nozzle tip and the grounded collector (or counter electrode). The magnitude of this distance affects the amount of solvent evaporated before any nanofibers are formed and deposited on the counter electrode. (5) *Environmental conditions*, such as humidity and temperature. A stable electrospinning process can be established through consideration and optimization of all of the above parameters. Post-treatment steps (for example calcination) are often employed after the collection of metal-oxide or ceramic nanofibers.[59,68,69] Such treatments can remove unnecessary polymer matrices in resultant fibers.

Usually, fibers are randomly distributed on the surface of the collector. In order to obtain aligned nanofibers, the basic electrospinning process must be modified. For example, pairs of split strips have been used to orient electrospun nanofibers.[70,71] In such a setup, the solid grounded counter electrode collector is substituted by a pair of conductive substrates with a void gap or patterned gold-electrode pairs on highly insulating substrates. When a continuous fiber is deposited on the collector, the electrostatic interaction in as-spun fibers initiates alignment. Another aligned nanofiber collection technique has been developed by Zussman's group.[72] They demonstrated the effectiveness of using a tapered and grounded wheel-like rotating bobbin as the collector to obtain parallel nanofibers. Since the collector is not flat, the challenge when using this bobbin collector is to transfer nanofibers onto a planar surface for possible integration with other devices and applications. Additional rotating collectors have been proposed by Kameoka and co-workers to define "scanning electrospinning" techniques.[73,74] A schematic of the process is provided in Figure 5.3(a). The grounded collector is connected to an electric rotational motor and rotates at high speed. A substrate is attached to the rotating collector where nanofibers will be deposited. Nanofibers can be deposited in an aligned fashion due to the rotation and relative motion between the electrospinning nozzle and the counter electrode. The orientation of nanofibers is controlled by the linear velocity of rotating collector. In this way, nanofibers can be aligned in a facile manner and collected on the flat substrate for further treatment and applications. Figure 5.3(b) and (c) show parallel-aligned ZrO_2 nanofibers on silicon wafers.

The basic electrospinning process has also been modified in a variety of ways to produce nanofibers with a range of secondary structures. For example, conjugated polymer nanofibers,[75] core-sheath nanofibers,[76-78] hollow nanofibers,[79,80] porous nanofibers[81] have all been produced by coaxial electrospinning from a two-fluid capillary spinneret. Nanofibers conjugated with fluorescent probes,[82] metal oxide particles[83] and nanoparticles of noble metals[84,85] have been reported.

For more detailed information about the electrospinning process, the reader is directed to some excellent review articles published elsewhere.[58,59,61,86,87] Theoretical analyses of the electrospinning process can also can be found.[88-91] However, due to complex effects of diversified processing parameters, accurate modelling of electrospinning is still challenging. In the following sections, we will focus on the apparatus used for coaxial electrospinning of silica nanofluidic tubes.

5.2.2.2 Nano-Scale Silica Fibers and Hollow Tubing Structures

Silica is common substrate material in biological applications due to its biocompatibility and excellent chemical and physical properties. Silica nanofibers can be successfully produced using the basic electrospinning set-up shown in Figure 5.1. An electrospinning solution can be prepared using the process developed by Kameoka and coworkers.[68] This solution is a blend of spin-on glass (SOG) intermediate coating IC1-200 solution and polyvinylpyrrolidone (PVP) (MW~1,300,000). SOG is the precursor of silica and PVP acts to introduce an appropriate viscosity to the precursor solution, which is required to extract a continuous jet and establish a stable electrospinning process. In its native form PVP is a white, hygroscopic powder which can be readily dissolved both in deionised water and in a large number of organic solvents. In the experiments described by Tsou,[92] the weight percentage of PVP in the whole solution is 4%. The mixture is stirred until the PVP is completely dissolved, resulting in a viscous and transparent solution. Subsequently the solution is dispensed into a glass syringe with a stainless steel needle as the electrospinning nozzle. The solution is pumped from the needle tip at a flow rate defined by an external syringe pump. The anode of a power supply is connected to the needle while the cathode is connected with a grounded collector (counter electrode), which is placed 5cm away from the needle tip. The applied voltage ranges from 3kV to 7kV. A piece of aluminium foil is attached on the collector to harvest silica fibers. After deposition, the fibers were calcined at 500°C for 5 hours to remove residual PVP and cross-linked silica matrices. After performing such a process, the morphologies of nanofiber samples with different process parameters were examined by SEM, as shown in Figure 5.4(a)-(d). It can be seen that high voltages lead to thinner fibers. This can be explained by noting that the electrostatic force becomes larger as the applied voltage increases. Hence, the liquid jet undergoes a higher stretching force. Moreover, as voltage increases, more electric charges are introduced into polymer solution. Other processing parameters can also be controlled to adjust the final diameter of silica nanofibers. These include SOG concentration and PVP concentration. Indeed, as a rule of thumb the higher the polymer concentration, the thicker the generated fiber.[92]

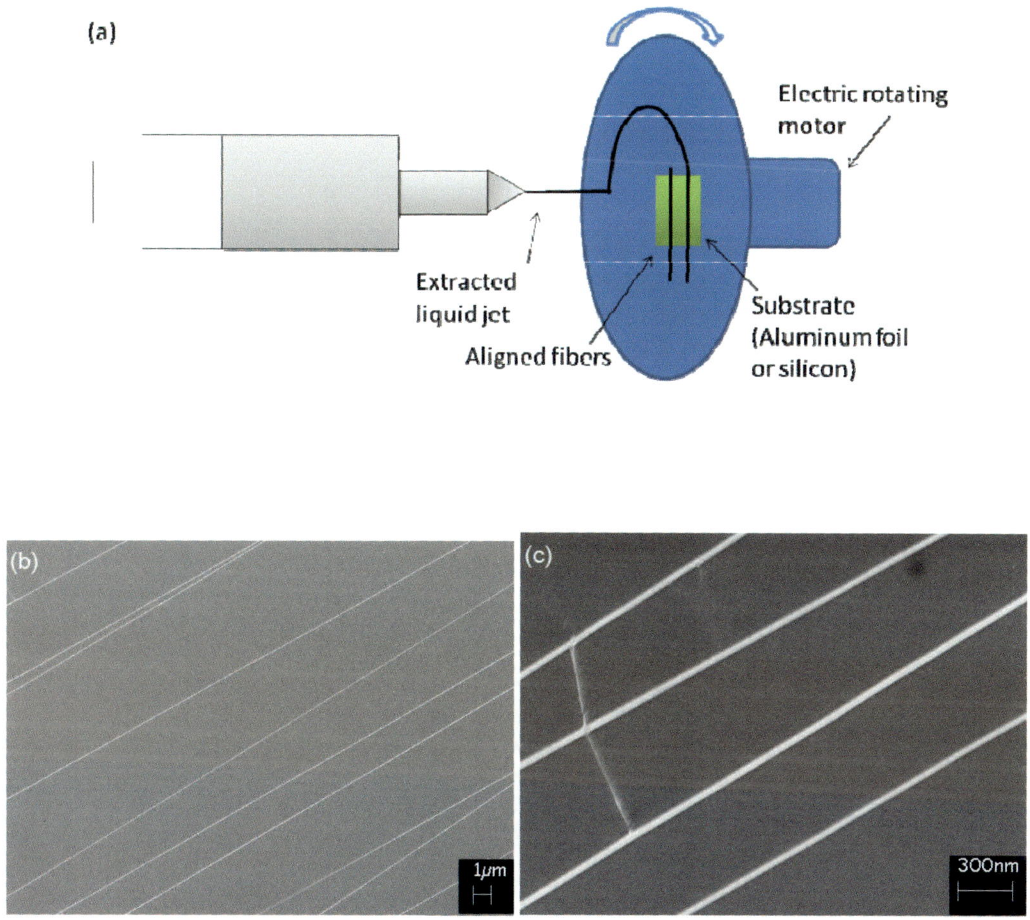

Figure 5.3 (a) Schematic diagram of scanning electrospinning set-up (b) and (c) SEM images of aligned ZrO_2 nanofibers fabricated by scanning electrospinning

The fabrication of hollow nanofluidic tubing by electrospinning has also been explored previously. Put simply, the methods are divided into two categories. In the first, nanofibers are used as templates, and in the second tubing fabrication occurs directly. In the former, coating materials are deposited on the surface of deposited nanofibers. Then the nanofiber's core is decomposed by heating or dissolution in an appropriate solvent to create the desired hollow structure. This method is conceptually similar to the fabrication of nanofluidic tubes by sacrificial layer methods discussed herein. The primary difference between these approaches lies in the generation of the sacrificial layer. Compared to other methods, electrospinning is more straightforward, economical and efficient. One of the key decisions made in this process is the choice of polymer to produce the nanofibers and shell materials. First of all, the polymer should be able to be electrospun and produce nanofibers with reduced dimensions. It should also be stable during the deposition process and easily removable in the final process. The shell material needs to be robust enough to retain its morphology during the decomposition of core and construction of the wall of nanofluidic tubing. To date, polymer, metal, metal oxide, and ceramic nanotubes have been produced using this general method.[93-97]

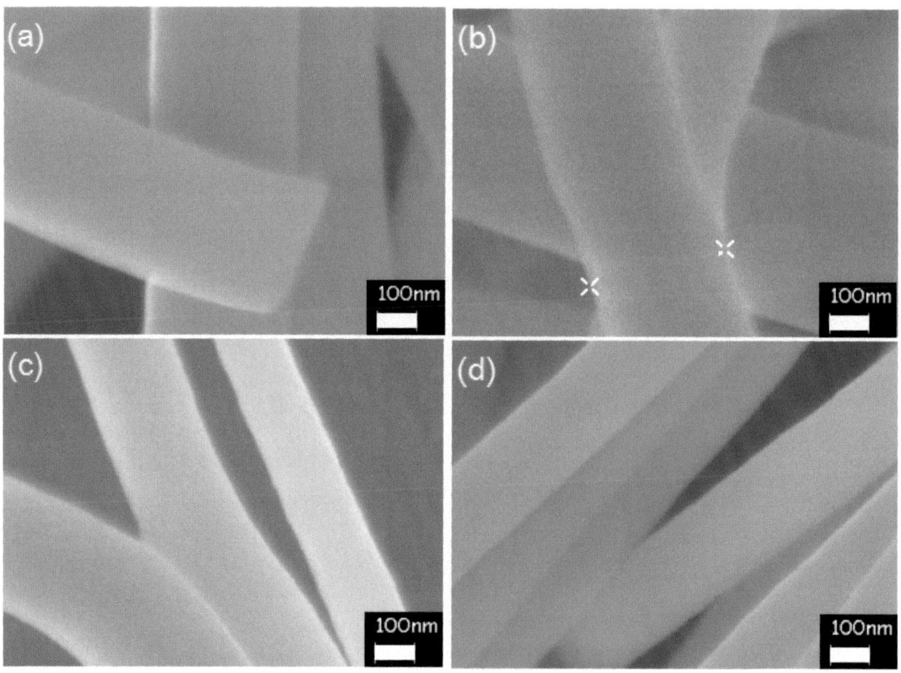

Figure 5.4 SEM images of silica nanofibers fabricated by electrospinning under a voltage of (a) 3kV (b) 4kV (c) 5kV (d) 7kV. Adapted from reference [92] and reproduced with permission.

The direct deposition method, which is typically called a coaxial electrospinning, defines the alternative approach for fabricating hollow nanofluidic tubing. For example, a two-capillary spinneret has been used to generate hollow tubing previously.[79-81] Moreover, a similar electrospinning deposition system was used to fabricate core-shell nanofibers.[76-78] By dispensing two immiscible materials (such as oil and water) in a coaxial spinneret, a coaxial jet can be extruded by electrostatic forces to create coaxial nanofibers. Subsequently, the core material can be removed by heating to create hollow nanotubes.

To simplify the fabrication process, an innovative coaxial electrospinning system incorporating a scanning function was developed by the precise alignment of two independent metal needles. Significantly, this system can easily generate oriented nanofluidic silica tubing.[98] Instead of using two-capillary nozzles as the coaxial source, two needles are aligned in an "L" shape arrangement. Using such an alignment, a coaxial jet can be generated directly. A schematic of this deposition system is shown in Figure 5.5(a).

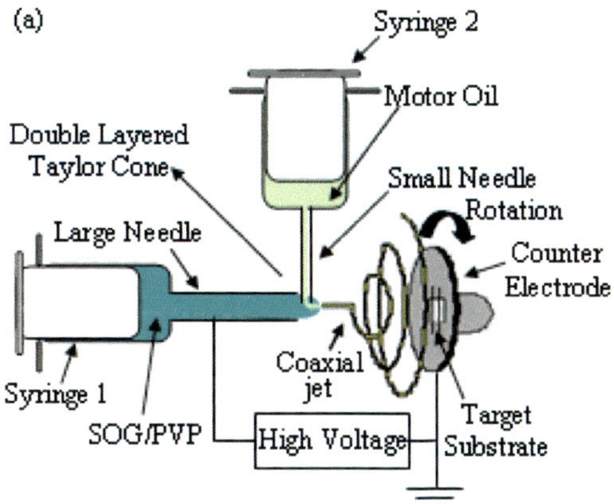

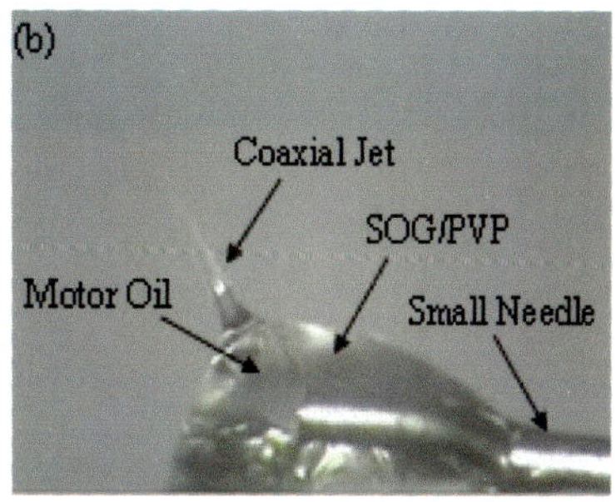

Figure 5.5 (a) Schematic diagram of coaxial electrospinning system (b) A double layered Taylor cone established on the needle tip. Adapted from M. Wang *et al.*, Applied Physics Letters, 2006, 88, 033106 and reproduced with permission.

Two syringes, syringe pumps, needles and positioning stages were used for this scanning coaxial electrospinning system. One syringe is filled with the shell material (a mixture of SOG and PVP prepared by the same procedures for the silica nanofiber fabrication) and dispensed through a large bore needle. The other syringe is infused with the core material, which is also the sacrificial material (SAE 10W30 motor oil), and connected to a small needle (33 Gauge, Hamilton). The SOG/PVP blend and motor oil are delivered at different rates by independent syringe pumps fixed on two positioning stages, and the anode of the power supply is connected to the large needle. By horizontal and vertical adjustment, the needle is precisely inserted in the Taylor cone of SOG/PVP stream. Due to their immiscibility, the core and shell phases do not inter-diffuse into each other and a double-layered Taylor cone is generated. The adjustment of needle position and the Taylor cone are monitored with a microscope. A live image of a core-shell Taylor cone is shown in Figure 5.5(b). The coaxial jet is pulled out and transported to the counter electrode which is rotating at a high speed.

To produce oriented nanofluidic tubes, scanning electrospinning is implemented. A piece of silicon wafer is attached onto the rotating counter electrode. Aligned nanofibers with SOG/PVP as the wall and motor oil as the core are deposited. Then, the wafer is calcined within a furnace at 850°C for 5 hours to cross link the silica as well as completely remove residual PVP and motor oil. In this way, hollow structured nanofibers are obtained. SEM images in Figure 5.6 confirm the effectiveness of this fabrication method. Nanofluidic tubes with a round/elliptical opening are shown in Figure 5.6(a) and Figure

5.6(b). A top view of the silicon wafer in Figure 5.6(c) shows the tubes aligned in a parallel fashion.

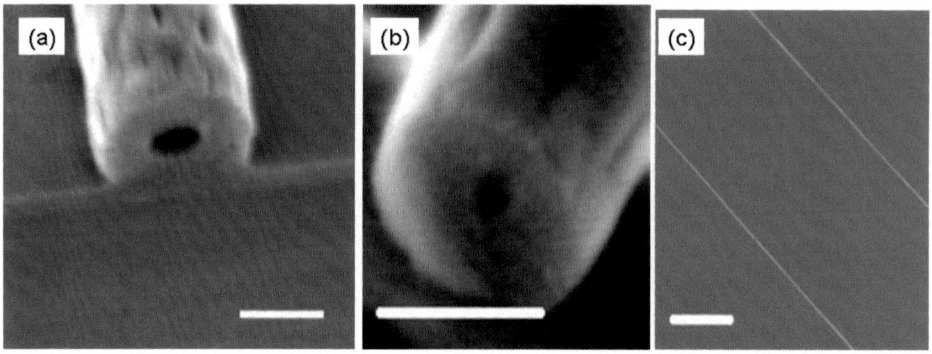

Figure 5.6 (a) SEM images of various cross sections of silica nanochannels and oriented parallel nanochannels. These nanochannels were deposited on silicon wafer and cleaved to show the cross sections. The scale bar is 100nm in (a) and (b), 10μm in (c). Adapted from M. Wang *et al.*, Applied Physics Letters, 2006, 88, 033106 and reproduced with permission.

5.2.2.3 Characterization of the Scanned Coaxial Electrospinning Process

When fabricating silica nanotubes, there are several process parameters that influence the morphology of nanofluidic tubes. These include the needle alignment, the flow rates of the SOG/PVP and motor oil and the viscosity of shell material. Precise alignment of two needles is necessary to ensure a stable coaxial source. The motor oil must be fully surrounded by SOG/PVP, which also means that the flow rate of motor oil cannot be too high so as to disrupt the interfacial layer of oil-SOG/PVP. If this happens, oil will spread out from the wrapping of the outer solution and disturb the spinning process. However, the flow rate of the motor oil must be high enough to allow formation of a continuous core. Inappropriate process parameters will have a significant and deleterious impact on the final structure of the spun fibers. Figure 5.7 illustrates the outcome of misalignment and improper flow conditions. Figure 5.7(a) describes a situation in which the oil-feeding needle tip is placed close to the edge of SOG/PVP Taylor cone. Although the motor oil can be extruded together with SOG/PVP solution and enclosed, it cannot be confined perfectly and firmly in the outer layer during the stretching, bending and whipping processes. Instead of forming an inner sacrificial medium, some oil molecules diffuse to the outer layer of the SOG/PVP fiber. Because of this diffusion, subsequent to calcination, an unsealed trench is created. Figure 5.7(b) shows a TEM image of discontinuous hollow tubes. In this example, the supply of oil is insufficient to cover the stretching of SOG/PVP fibers under electrostatic force. Therefore, tubular segments of nanofibers are formed. Unenclosed nanofiber segments can also be seen along with a non-uniformity in nanofiber diameter (Figure 5.7(c)). This is likely to be a result of both misalignment and inadequate flow rate.

The flow rate of the oil stream also has influence on the inner diameter of the tubing. As the flow rate increases within a reasonable range, more core material is pumped out per unit time, resulting in a larger core and inner diameter. An oil flow rate between

2μl/min and 6μl/min defines an appropriate range for generating nanoscale silica tubes for 5 wt% PVP in an SOG polymer solution with an average electric field of 525V/cm. It should be noted that lower and higher flow rates may also work, but may require adjustment of other processing parameters, such as PVP concentration, electric field and needle size.

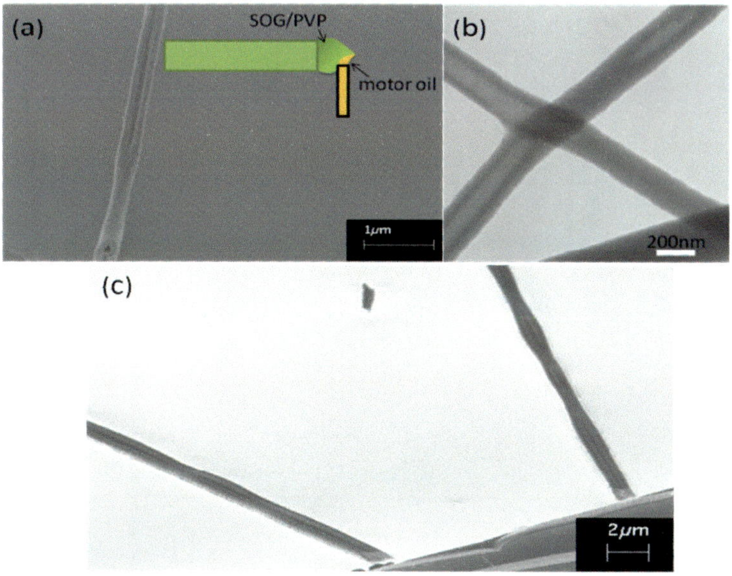

Figure 5.7 (a) and (c) SEM images of incomplete nanofluidic structure (b) TEM image of incomplete nanofluidic structure.

The concentration of PVP has a direct influence on the viscosity of the polymer solution. A typical range of PVP concentration for the described experiments is between 4 wt. % and 8 wt. %. As the PVP concentration increases in the solution, a concomitant increase in viscous stress provides better confinement of the oil phase. Accordingly, the oil can be extruded into a thinner core, thus generating smaller diameter fluidic tubing. The nano-tubing shown in Figure 5.6(a) has an average of 60nm and was fabricated with a 5 wt. % PVP in SOG solution. Conversely, the tubing shown in Figure 5.6(b) has an average diameter of 15nm and was obtained from an 8 wt. % PVP in SOG solution.

As the concentration of PVP increases, the flow rate of motor oil must be increased accordingly to compensate for the increased viscous stress caused by the increased PVP concentration. By balancing all these process parameters, silica nanofluidic tubing can be created successfully (Figure 5.6(a) and (b)). From SEM image analysis, some broken (unenclosed) nanofluidic tubes can be seen. One example is shown in Figure 5.8(a). In this image, a round channel shape is clearly observed. This confirms that the oil is extruded out as the core and retains a cylindrical structure without collapsing in the electrospinning and calcination process. Through use of proper process parameters, continuous hollow nanofluidic tubes can be created. A TEM image in Figure 5.8(b) demonstrates such an example. The brighter part of this image illustrates the hollow interior with the darker region representing the wall.

Previous research on controlling the diameter of electrospun nanofibers, has suggested that the nanofiber diameter can be controlled by varying the deposition distance. This is because the transit time for the electrified jet increases correspondingly as the deposition distance increases. With a longer time in transit, more solvent evaporates and the diameter of nanofiber can be reduced. Our group has also studied how the deposition distance affects the dimensions of nanofluidic tubes. Figure 5.9 illustrates the variation of the inner diameter (I.D.) and outer diameter (O.D) of nanochannels as a function of deposition distance. It indicates that no obvious change in I.D. and O.D. occurs when the deposition distance varies. This observation might be explained by the structure of coaxial liquid jets extruded during the electrospinning process. The motor oil cannot evaporate through the SOG/PVP wall layer. Thus, the shape of motor oil is maintained and the I.D. of the nanofluidic tube will not vary. Additionally, the O.D of nanofluidic tubes shows no obvious dependence on the deposition distance in a range of 5.3cm to 15.3cm.

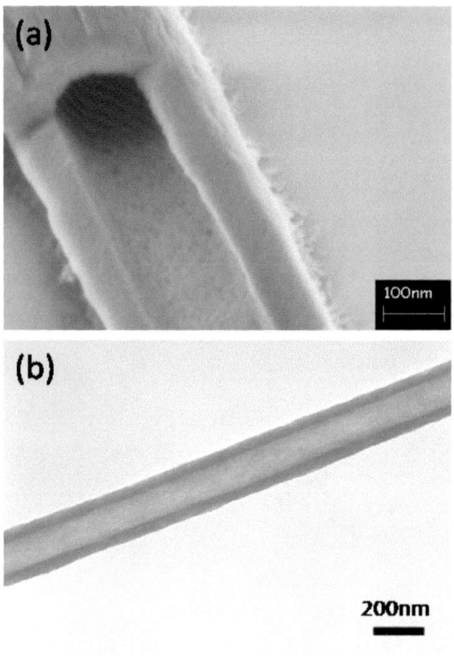

Figure 5.8 (a) SEM image of broken nanofluidic tubings. (b) TEM image of continuous hollow nanotube.

To control the orientation of the coaxial nanofibers, we have used the scanning technique to create parallel nanofibers. A rotating counter electrode with a linear velocity between 660 and 850cm/s was placed at the collecting position, and a relative motion between the coaxial jet and target substrate (a silicon wafer attached to the counter electrode 4.5cm away from the centre) was used. The orientation of nanofibers can be controlled via a linear scanning motion rather than the random spinning motion.[73] In conclusion, silica nanofluidic tubes can be fabricated using the scanning, coaxial electrospinning process. The instrumentation can be set up easily and involves the use of two aligned syringes rather than specially designed spinnerets. By rotating the counter electrode and removing the core of the coaxial nanofibers, parallel oriented and circular silica nanofluidic tubes with an I.D. as small as 15 nm can be formed. Due to their extremely small dimensions, these nanochannels have many potential applications, especially in facilitating high-efficiency single molecule detection. This will be discussed in the proceeding paragraphs.

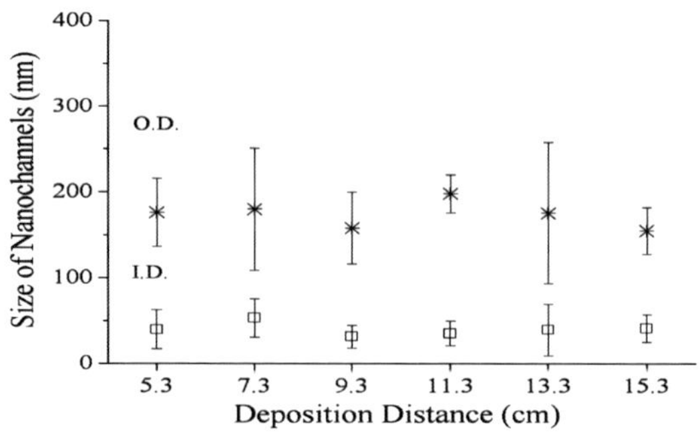

Figure 5.9 The inner (I.D.) and outer (O.D.) dimensions of nanofluidic tubings as the function of the deposition distance.

5.2.3 Heat-Induced Stretching Method

Due to the physical properties of silica, the thermal deformation of micro-silica tubes is another way to form nanoscale tubing. Indeed, there is a long history of heat deformation of glass in industrial and artistic applications. Moreover, a detailed theoretical analysis of thermally-induced glass tube stretching for fabricating glass microelectrodes can be found elsewhere.[99] Non-silica based glass has also been shown to form holey fibers by thermally-induced drawing methods.[100] However, the fabrication of nanotubes by this approach has not been investigated in any detail. A prototype system for nanotube fabrication has been recently developed in our group and is shown in Figure 5.10 (unpublished work). The primary parts of this system include two stages, a butane torch, capillary tubing with a 12μm thick coating of polyimide and a fused silica capillary cutter. Stage 1 is fixed onto the lab bench, and Stage 2 can be mechanically moved. A void gap of ~1cm is created between the edges of the two stages. The ends of the capillaries are glued to the each of the two stages. A butane torch is then placed in the gap beneath the tubing. After the torch is ignited the polyimide coating can be rapidly eliminated in a defined area. Within a further few seconds, the fused silica is softened by heat. Stretching of glass tubing process is performed using Stage 2, which is shifted in the x-direction. The fused silica tubing gradually deforms, tapers and shrinks. The inset of Figure 5.10 shows a magnified schematic diagram of a deformed part of the tubing under the application of heat.

The shift distance of Stage 2 has a significant effect on the size of final tubing. The final tubing may be cut with a razor blade in the centre to show the cross section. Figure 5.11(a)-(d) show the cross section of stretched fused silica tubes with a shift of 1cm, 2cm, 3cm and 5cm respectively. With a 1cm shift, the inner diameter of silica tubing is reduced to ~250nm. For longer shift distances, e.g. 5cm, the inner diameter is reduced but finally collapses under heating and results in a solid segment of fused silica. The movement of Stage 2 is associated with the stretching force to elongate the tube. As Stage 2 moves to the desired point, the butane torch is removed allowing the tubing to cool down and solidify. The ends of the tubing are released and the final tubing can be picked up on a glass slide or silicon wafer for further use. The two ends still retain their shape since they are fixed on the stage and kept cool. The centre part of tubing, in the heating zone, is deformed, and the inner diameter can be reduced to submicron dimensions. However, it should be noted that the diameter is not uniform along the capillary (see inset of Figure 5.10). Importantly, the micro-scale ends can be easily integrated with other components such as reservoirs for feeding in sample solution.

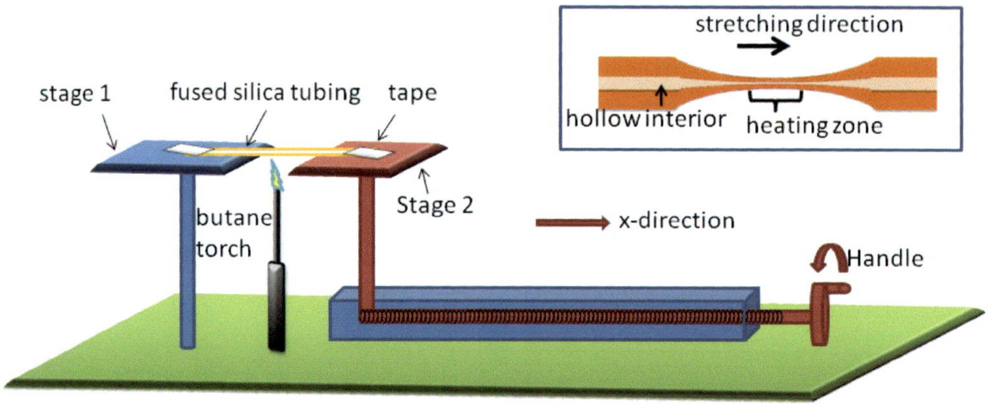

Figure 5.10 Schematic of a prototype for heat-induced stretching method

In single molecule detection experiments, plastic reservoirs were glued over the tubing ends with 5-minute epoxy. To assess their structural integrity, a 4.9μM solution of 5-Iodoacetamidofluorescein (IAF) in a pH 9 EPPS buffer was injected into the reservoir. The solution was infused into the tubing by capillary force. The tubes were then monitored using fluorescence microscopy. Under normal conditions the sample solution flows into the tubing very quickly. However, the linear velocity decreases dramatically as the solution approaches the centre of the tube where the diameter is the smallest (and sub-micron). The image shown in Figure 5.12 illustrates the IAF-filled fused silica tubing with a minimum inner diameter of approximately 300nm. This simple visualisation experiment confirms that the heat-induced stretching method can create a continuous hollow nanofluidic tube directly from a micro-scale tube. The advantage of this fabrication method is its simplicity and low cost. Indeed, high throughput fabrication can be achieved if multiple tubes are fixed and stretched together. The primary drawback associated with the technique is related to the fact that the tube dimension is not uniform along the entire length of the conduit. Furthermore, it is difficult to fabricate nanofluidic channels with an internal diameter less than 150 nm.

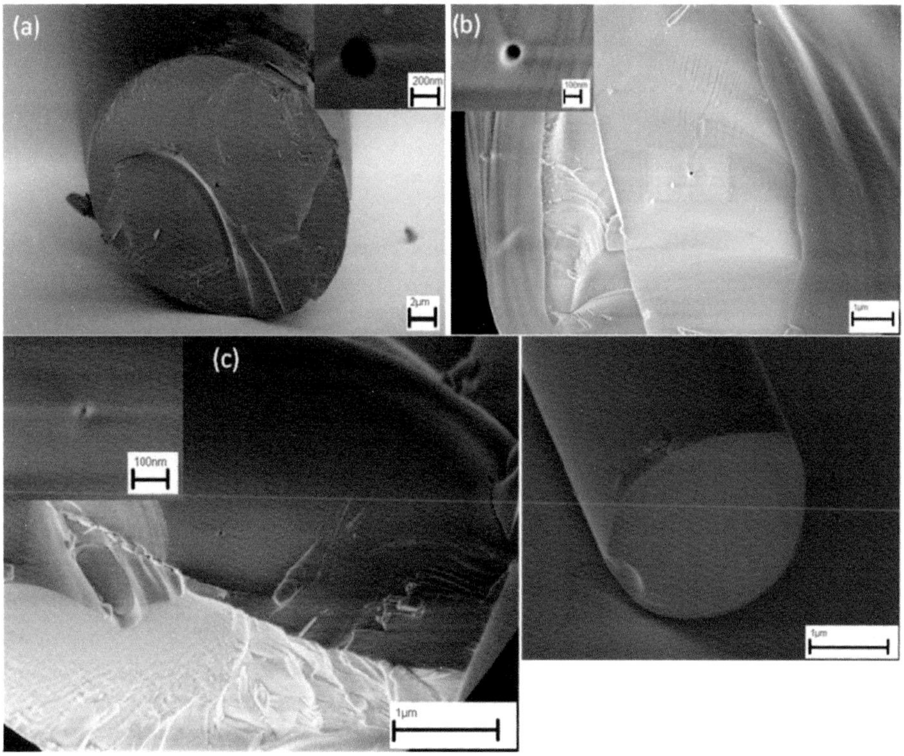

Figure 5.11 SEM images of stretched fused silica nanofluidic tubings with shift of (a)1cm (b) 2cm (c) 3cm (d) 5cm. (Jun Kameoka, unpublished work)

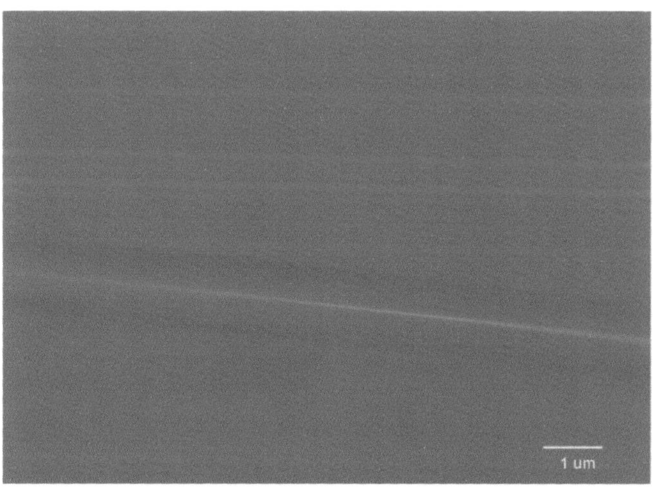

Figure 5.12 Fluorescence image of IAF-filled fused silica tubing. (Jun Kameoka, unpublished work)

5.3 ANALYSIS OF SINGLE MOLECULES USING NANOFLUIDIC TUBES

5.3.1 Experimental Setup

To investigate the performance of nanofluidic tubing for single molecule analysis, IAF was utilized as a sample solution. A schematic diagram of the detection system employed is illustrated in Figure 5.13. The scanning coaxial electrospinning technique was used to fabricate silica nanofluidic tubes as previously described. A substrate is used to hold and align the electrospun tubes. The substrate needs to be both robust (in order to withstand the high-temperature calcination process) and also transparent to visible light (to allow efficient molecule excitation and fluorescence detection). Accordingly, a quartz wafer is commonly used to collect aligned electrospun tubes. Plastic reservoirs can be attached at the both sides of nanofluidic tubes, and the sample solution injected into a reservoir and infused into the tubing using capillary force. For the experiments described herein an 800 nm laser with a pulse duration of 80 fs and a repetition rate of 80 MHz was focused into the nanofluidic tubing. As fluorescent molecules pass through the detection volume, they are excited (via two-photon absorption) and emit fluorescence photons in the visible region of the electromagnetic spectrum. Fluorescence photons are collected and detected using an avalanche photodiode, and are then processed and displayed as photon burst counts per unit time.

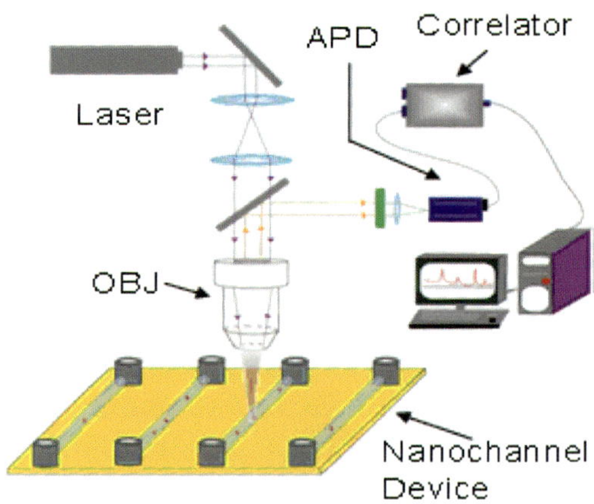

Figure 5.13 Schematic setup for single molecule detection (OBJ = microscope objective, APD = avalanche photodiode)

5.3.2 Detection and Measurement of Single Molecules in Nanofluidic Channels

Single molecule detection has been performed within electrospun nanofluidic tubes having an inner diameter of 25 nm. A 4.9 μM solution of IAF in EPPS buffer (pH=9) was infused into the nanotubes. Figure 5.14(a) shows an image of aligned IAF filled nanotubes. Channel integrity is confirmed through the observation that no dye leaks from the nanotubes. Single molecule detection was then demonstrated within this nanotubing. Figure 5.14(c) and (d) display the photon burst signals before and after filling the nanofluidic tubes with IAF molecules. The *x*-axis defines the time points at which data are acquired during analysis, with the *y*-axis representing the number of photon counts per bin time. In simple terms, a sub-femtolitre probe volume is defined by the overlap between a focused laser beam (near the diffraction limit) and the nanofluidic channel. As a single molecule diffuses through this volume it can be repeatedly cycled between its ground and excited states to yield a burst of fluorescence photons which can be collected and detected. Figure 5.14(c) and (d) report photon counts every 10μs bin within a 2ms time period. For an empty nanofluidic tube, no photons are observed. After the injection of IAF molecules, photon burst peaks originating from single molecule events are detected. Based on the size of nanotubing and the size of the laser beam waist, the detection volume of this system was calculated to be 150 zeptolitres. With a 4.9μM solution, the average number of molecules in the detection volume at any instant is approximately 0.4, which satisfies the single molecule standard. To put this in context, for a microfluidic channel with a 1μm inner diameter, the average number of molecules in the detection volume for a 4.9μM solution is approximately 695. This occupancy is clearly far from the requirements of single molecule detection! Indeed, for this microfluidic environment, single molecule detection can only be achieved with a solution concentration of less than 7 nM.

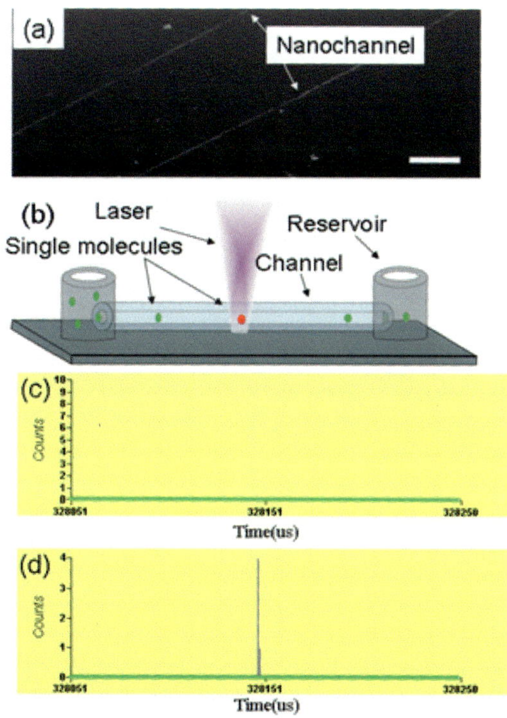

Figure 5.14 (a) Fluorescent optical micrograph of aligned nanochannel filled with IAF. The scale bar is 10 μm. (b) Schematic diagram of single molecule detection. (c) Photon counts of a blank nanochannel. (d) Photon counts of a nanochannel filled with a 4.9 μM of IAF solution. Adapted from M. Wang *et al.*, Applied Physics Letters, 2006, 88, 033106 and reproduced with permission.

5.3.3 Electrokinetic Molecule Transport in Nanofluidic Tubing

Molecule transport in the photon counting experiments described above relies on capillary force. However, this force is transient and once the channel is filled with solution, molecular transport by capillary forces is replaced by molecular diffusion. To overcome diffusional-induced molecule transport, electrokinetic forces have been employed to transport molecules. On application of an electric potential between the two reservoirs charged ions in the solutions are motivated by the electrophoretic effect, whilst neutral molecules move due to electroosmosis.

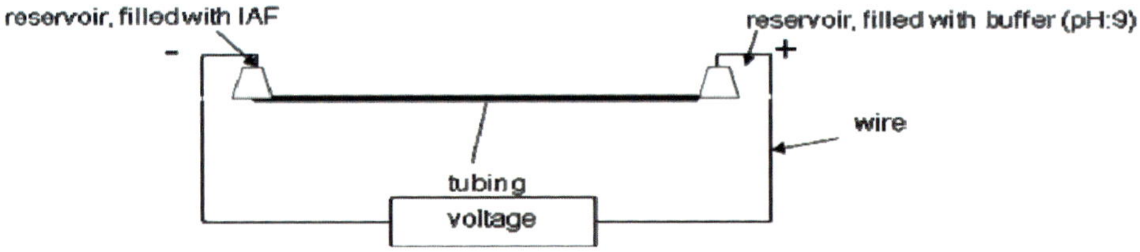

Figure 5.15 Schematic diagram of experiment setup for electrokinetic effect in nanofluidic tubing.

The IAF-filled fused-silica nanofluidic tubing (200 nm diameter) was used to demonstrate the electrokinetic transport of molecules. A schematic diagram of the set-up is shown in Figure 5.15. Two plastic reservoirs were glued at the ends of tube, and a 4.9 μM IAF solution was introduced into the left reservoir for 2 hours until the entire device was filled up with the sample solution. Then EPPS buffer was injected into the right reservoir and an electric field applied across the tube. The excitation laser was focused into the tubing, and photon burst counts monitored as a function of applied voltage.

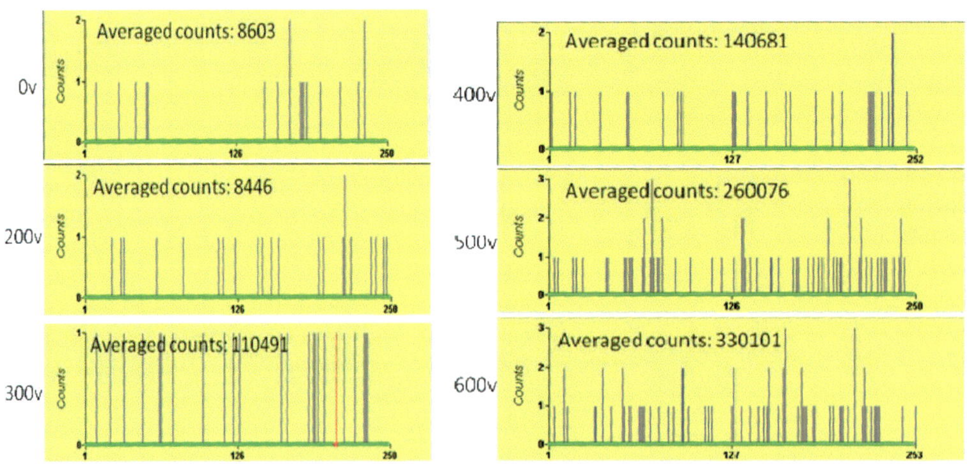

Figure 5.16 Photon burst signal monitored across a nanofluidic tube under different voltages.

Figure 5.16 shows the results of this investigation. The averaged counts represent the total photon counts collected over a period of 10 seconds. It can be observed that this number increases dramatically as a function of applied voltage, meaning that more IAF molecules pass through the detection volume within a given period of time. Thus, the transport of molecules in nanofluidic tubing can be controlled by an electric field.

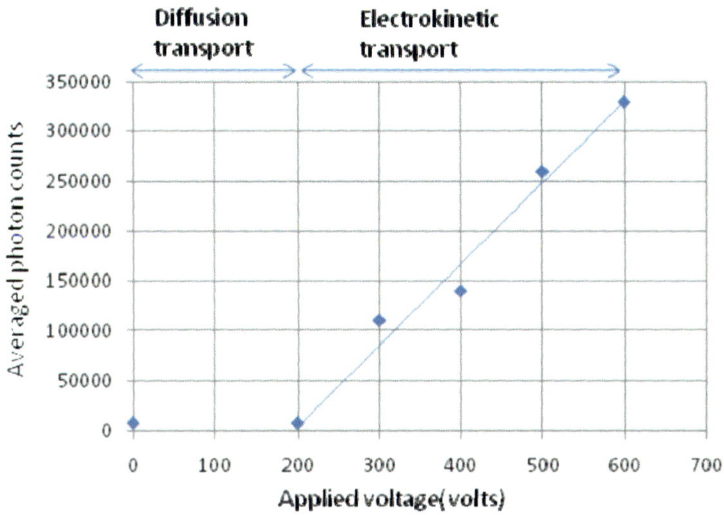

Figure 5.17 Averaged photon counts (collected over a period of 10 seconds) for a 4.9 μM IAF solution in fused-silica nanofluidic tubing versus applied voltage.

A plot of averaged photon counts as a function of applied voltage is shown in Figure 5.17. Within the range of 0-200 volts, diffusion dominates molecular transport. As the applied voltage exceeds 200 volts, the electrokinetic force begins to dominate. Indeed, between 200 and 600 volts, the averaged photon counts exhibit a quasi-linear increasing relationship with the applied voltage. Molecular diffusion is still occurring but the electrokinetic force turns is now the main dragging force on target molecules. As the applied potential increases, the relative contribution of the electrokinetic force becomes larger and larger and diffusional effects can be neglected above 500 volts.

5.4 CONCLUSIONS

In this chapter, two different nanofluidic tube fabrication techniques, scanned coaxial electrospinning and heat-induced stretching, have been presented in detail. The importance of various process parameters on the efficacy of each approach has been discussed. Using either method, circular nanofluidic tubes can be produced. Such conduits have been shown to allow for electrokinetic pumping of small molecules. To demonstrate the merits of using nanofluidic tubes for single molecule analysis, photon bursts originating from single 5-Iodoacetamido-fluorescein molecules moving through nanofluidic tubes has been reported. The key benefit of using nanofluidic tubing in single molecule detection applications lies in the ability to define extremely small detection probe volumes, and thus enable the detection of single molecules in samples of high analyte concentration. Electrokinetic molecule transport in nanofluidic tubing has also been investigated. From the photon burst count analysis, it can be concluded that molecule transport is enhanced in nanofluidic tubing under electrokinetic potential. The potential application of such nanofluidic tubing in identifying extremely low concentrations of target molecules among a high concentration of background analyte is particularly appealing in diagnostic applications where the detection of tumour markers in bodily fluids will undoubtedly define novel approaches for the early diagnosis of cancers and other diseases.

5.5 ACKNOWLEDGEMENTS

This research was supported by the Texas A&M University Start-up Fund. Photon burst experiments were performed at the Department of Biochemistry and Biophysics in Texas A&M University. Support from Professor Gregory D. Reinhart and Andrew N. Bigley is gratefully acknowledged. We would also like to thank our group members, Pei-Hsiang Tsou and Nan Jing for insightful discussions and acquisition of some of the SEM/TEM images. The FE-SEM acquisition was supported by the National Science Foundation under Grant No. DBI-0116835.

References

1. M. Foquet, J. Korlach, W. Zipfel, W. W. Webb and H. G. Craighead, *Anal. Chem.*, 2002, **74**, 1415.
2. M. Foquet, J. Korlach, W. R. Zipfel, W. W. Webb and H. G. Craighead, *Anal. Chem.*, 2004, **76**, 1618.
3. S. M. Stavis, J. B. Edel, K. T. Samiee and H. G. Craighead, *Lab Chip*, 2005, **5**, 337.
4. B. Valeur, *Molecular Fluorescence*, Wiley-VCH, 2002.
5. S. R. Aragon and R. Pecora, *Biopolymer*, 1975, **14**, 119.
6. J. P. Bouchaud and A. Georges, *Phys. Rep*, 1990, **195**, 127.
7. W. Denk, J. J. Strickler and W. W. Webb, *Science*, 1990, **248**, 73.
8. L. Edman, U. Mets and R. Rigler, *Proc. Nat. Accd. Sci.*, 1996, **93**, 6710.
9. M. Ehrenberg and R. Rigler, *Chem. Phy.*, 1974, **4**, 390.
10. M. Ehrenberg and R. Rigler, *Quart. Rev. Biophys.*, 1976, **9**, 69(61).
11. M. Brinkmeir, K. Dorre, J. Stephan and M. Eigen, *Anal. Chem.*, 1999, **71**, 609.
12. Y. Chen, J. D. Muller, S. Y. Tetin, J. D. Tyner and E. Gratton, *Biophy. J.*, 2000, **79**, 1074.
13. P. Dittrich, F. Campeggi, M. Jahnz and P. Schwile, *Biol. Chem*, 2000, **382**, 491.
14. C. Eggeling, J. R. Fries, L. Brand and C. A. M. Seidel, *Proc. Nat. Accd. Sci. USA*, 1998, **95**, 1556.
15. M. Eigen and R. Rigler, *Proc. Nat. Accd. Sci. USA*, 1994, **91**, 5740.
16. E. L. Elson and D. Magde, *Biopolymers*, 1974, **13**, 1.
17. A. Gennerich and D. Schild, *Biophys. J.*, 2000, **79**, 3294.
18. U. Haupts, S. Maiti, P. Scheille and W. W. Web, *Proc. Nat. Accd. Sci. USA*, 1998, **95**, 13573.

19. K. G. Heinze, A. Kolterman and P. Schwille, *Proc. Nat. Accd. Sci. USA*, 2000, **97**, 10377.
20. P. Kask, P. Piksarv, U. Mets and E. Lippma, *Eur. Biophys. J.*, 1987, **14**, 257.
21. M. Kinjo and R. Rigler, *Nucleic Acids Res.*, 1995, **23**, 1795.
22. J. Korlach, P. Schwille and W. W. Webb, *Proc. Nat. Accd. Sci. USA*, 1999, **95**, 1421.
23. U. Kettling, A. Kolterman, P. Schwille and M. Eigen, *Proc. Nat. Accd. Sci. USA*, 1998, **95**, 14116.
24. D. Magde, E. L. Elson and W. W. Webb, *Biopolymers*, 1974, **13**, 29.
25. M. J. Levene, J. Korlach, S. W. Turner, M. Foquet, H. G. Craighead and W. W. Webb, *Science*, 2003, **299**, 682.
26. L. Sacconi, D. A. Dombeck and W. W. Webb, *PNAS*, 2006, **103**, 3124.
27. P. Schwille, S. Kummer, A. A. Heikal, W. E. Moerner and W. W. Webb, *PNAS*, 2000, **97**, 151.
28. S. T. Hess, S. Huang, A. A. Heikal and W. W. Webb, *Biochemistry*, 2002, **41**.
29. J. Khandurina, S. C. Jacobson, L. C. Waters, R. S. Foote and J. M. Ramsey, *Anal. Chem.*, 1999, **71**, 1815.
30. E. T. Lagally, P. C. Simpson and R. A. Mathies, *Sensors and Actuators B.*, 2000, **63**, 138.
31. M. Foquet, J. Korlach, W. Zipfel, W. W. Webb and H. G. Craighead, *Anal. Chem.*, 2002, **74**, 1415.
32. M. Foquet, J. Korlach, W. R. Zipfel, W. W. Webb and H. G. Craighead, *Anal. Chem.*, 2004, **76**, 1618.
33. C. Y. Kung, M. D. Barnes, N. Lermer, W. B. Whitten and J. M. Ramsey, *Anal. Chem.*, 1998, **70**, 658.
34. R. Qiao and N. R. Aluru, *Appl. Phys. Lett.*, 2005, **86**, 143105.
35. C. W. Hollars and J. Puls, *Anal. Bioanal. Chem*, 2006, **385**, 1384.
36. M. A. Stavis and J. B. Edel, *Lab Chip*, 2005, **5**, 337.
37. A. L. Garcia, L. K. Ista, D. M. Petsev, M. J. O'Brian, P. Bisong, A. A. Mammoli, S. R. J. Brueck and G. P. Lopez, *Lab Chip*, 2005, **5**, 1271.
38. A. Plecis, R. B. Schoc and P. Renaud, *Nano Lett.*, 2005, **5**, 1147.
39. S. Pennathur and J. G. Santiago, *Anal. Chem.*, 2005, **77**, 6772.
40. G. Karniadakis, A. Beskok and A. Aluru, *Microflows and Nanoflows*, 2007.
41. A. Rasmussen, M. Gaitan, L. E. Locascio and M. E. Zaghloul, *J. Microelectromech. Syst.*, 2001, **10**, 286.
42. D. Mijatovic, J. C. T. Eijkel and A. v. d. Berg, *Lab Chip*, 2005, **5**, 492.
43. J. Haneveld, H. Jansen, E. Berenschot, N. Tas and M. Elwenspoek, *J. Micromech. Microengin.*, 2003, **13**, S62.
44. E. D. Fabrizio, R. Fillipo, S. Cabrini, R. Kumar, F. Perennes, M. Altissimo, L. Businaro, D. Cojac, L. Vaccari, M. Prasciolu and P. Candeloro1, *J. Phys.: Condens. Matter*, 2004, **16**, S3517.
45. T. K. K. Matsumoto, M. Nakao, Y. Hatamura, T. Kitamori, T. Sawada, in *Proc. IEEE MEMS*, 1998, p. 127.
46. C. K. Harnett, G. W. Coates and H. G. Craighead, *J. Vac. Sci. Technol. B.*, 2001, **19**, 2842.
47. K. A. Mahabadi, I. Rodriguez, S. C. Haur, J. A. v. Kan, A. A. Bettiol and F. Watt, *J. Micromech. Microeng.*, 2006, **16**, 1170.
48. D. M. Cannon, B. R. Flachsbart, M. A. Shannon and J. V. Sweedler, *Appl. Phys. Lett.*, 2004, **85**, 1241.
49. S. Arscott and D. Troadec, *Nanotechnology*, 2005, **16**, 2295.

50. M. J. O'Brien, P. Bisong, L. K. Ista, E. M. Rabinovich, A. L. Garcia, S. S. Sibbett, G. P. Lopez and S. R. J. Brueck, *J. Vac. Sci. Technol. B.*, 2003, **21**, 2941.
51. H. Cao, Z. Yu, J. Wang, J. O. Tegenfeldt, R. H. Austin, E. Chen, W. Wu and S. Y. Chou, *Appl. Phys. Lett.*, 2002, **81**, 174.
52. L. J. Guo, X. Cheng and C. Chou, *Nano Lett.*, 2003, **4**, 69.
53. H. H. Solak, C. David, J. Gobrecht, L. Wang and F. Cerrina, *J. Vac. Sci. & Tech. B: Microelectronics and Nanometer Structures*, 2002, **20**, 2844.
54. S. H. Zaidi and S. R. J. Brueck, *J. Vac. Sci. & Tech. B: Microelectronics and Nanometer Structures*, 1993, **11**, 658.
55. M. Wang, N. Jing, I.-H. Chou, G. L. Cote and J. Kameoka, *Lab Chip*, 2007, **7**, 630.
56. J. Han and H. G. Craighead, *J. Vac. Sci. Technol. A.*, 1999, **17**, 2142.
57. W. Li, J. O. Tegenfeldt, L. Chen, R. H. Austin, S. Y. Chou, P. A. Kohl, J. Krotine and J. C. Sturm, *Nanotechnology*, 2003, **14**, 578.
58. R. Ramaseshan, S. Sundarrajan, R. Jose and S. Ramakrishna, *J. Appl. Phys.*, 2007, **102**, 111101.
59. D. Li, J. T. McCann and Y. Xia, *J. Am. Ceram. Soc.*, 2006, **89**, 1861.
60. A. Formhals, 1975504, US Patent, 1934.
61. D. Li and Y. Xia, *Adv. Mater.*, 2004, **16**, 1151.
62. S.-S. Choi, Y. S. Lee, C. W. Joo, S. G. Lee, J. K. Park and K.-S. Han, *Electrochimica Acta* 2004, **50**, 339.
63. S. W. Choi, S. M. Jo, W. S. Lee and Y. R. Kim, *Adv. Mater.*, 2003, **15**, 2027.
64. K. H. Lee, H. Y. Kim, M. S. Khil, Y. M. Ra and D. R. Lee, *Polymer*, 2003, **44**, 1287.
65. S. A. Theron, E. Zussman and A. L. Yarin, *Polymer*, 2004, **45**, 2017.
66. Y. You, B.-M. Min, S. J. Lee, T. S. Lee and W. H. Park, *J. Appl. Polymer Sci.*, 2004, **95** 193
67. C. L. Casper, J. S. Stephens, N. G. Tassi, D. B. Chase and J. F. Rabolt, *Macromolecules*, 2004, **37** 573.
68. J. Kameoka, S. S. Verbridge, H. Liu, D. A. Czaplewski and H. G. Craighead, *Nano Lett.*, 2004, **4**, 2105.
69. D. Li, T. Herricks and Y. Xia, *Appl. Phys. Lett.*, 2003, **83**, 4586.
70. D. Li, Y. Wang and Y. Xia, *Nano Lett.*, 2003, **3**, 1167.
71. D. Li, Y. Wang and Y. Xia, *Adv. Mater.*, 2004, **16**, 361.
72. A. Theron, E. Zussman and A. L. Yarin, *Nanotechnology*, 2001, **12**, 384.
73. J. Kameoka and H. G. Craighead, *Appl. Phys. Lett.*, 2003, **83**, 371.
74. J. Kameoka, R. Orth, Y. Yang, D. Czaplewski, R. Mathers, G. W. Coates and H. G. Craighead, *Nanotechnology*, 2003, **14**, 1124.
75. D. Li, A. Babel, S. A. Jenekhe and Y. Xia, *Adv. Mater.*, 2004, **16**, 2062.
76. Z. Sun, E. Zussman, A. L. Yarin, J. H. Wendorff and A. Greiner, *Adv. Mater.*, 2003, **15**, 1929.
77. J. H. Yu, S. V. Fridrikh and G. C. Rutledge, *Adv. Mater.*, 2004, **16**, 1562.
78. Y. Zhang, Z.-M. Huang, X. Xu, C. T. Lim and S. Ramakrishna, *Chem. Mater.*, 2004, **16**, 3406.
79. D. Li, J. T. McCann and Y. Xia, *Small*, 2005, **1**, 83.
80. D. Li and Y. Xia, *Nano Lett.*, 2004, **4**, 933.
81. J. T. McCann, D. Li and Y. Xia, *J. Mater. Chem*, 2005, **15**, 735.
82. X. Wang, Y.-G. Kim, C. Drew, B.-C. Ku, J. Kumar and L. A. Samuelson, *Nano Lett.*, 2004, **4**, 331.
83. C. Drew, X. Liu, D. Ziegler, X. Wang, F. F. Bruno, J. Whitten, L. A. Samuelson and J. Kumar, *Nano Lett.*, 2003, **3**, 143.

84. G.-M. Kim, A. Wutzler, H.-J. Radusch, G. H. Michler, P. Simon, R. A. Sperling and W. J. Parak, *Chem. Mater.*, 2005, **17**, 4949.
85. A. C. Patel, S. Li, C. Wang, W. Zhang and Y. Wei, *Chem. Mater.*, 2007, **19**, 1231.
86. A. Greiner and J. H. Wendorff, *Angew. Chem. Int. Ed.*, 2007, **46**, 5670.
87. T. Subbiah, G. S. Bhat, R. W. Tock, S. Parameswaran and S. S. Ramkumar, *Journal of Applied Polymer Science*, 2005, **96**.
88. M. M. Hohman, M. Shin, G. Rutledge and M. P. Brenner, *Phys. Fluids*, 2001, **13**, 2201.
89. M. M. Hohman, M. Shin, G. Rutledge and M. P. Brenner, *Phys. Fluids*, 2001, **13**, 2221.
90. Y. M. Shin, M. M. Hohman, M. P. Brenner and G. C. Rutledge, *Polymer*, 2001, **42**, 9955.
91. A. L. Yarin, S. Koombhongse and D. H. Reneker, *J. Appl. Phys.*, 2001, **90**, 4836.
92. P.-H. Tsou, in *Electrical Engineering*, Texas A&M University, College Station, Electrospinning of Silica Nanofibers: Characterization and Application to Biosensing, Thesis for Master of Science, 2006.
93. M. Bognitzki, H. Hou, M. Ishaquc, T. Frese, M. Hellwig, C. Schwarte, A. Schaper, J. H. Wendorff and A. Greiner, *Adv. Mater.*, 2000, **12**, 637.
94. R. A. Caruso, J. H. Schattka and A. Greiner, *Adv. Mater.*, 2001, **13**, 1577.
95. D. A. Czaplewski, J. Kameoka, R. Mathers, G. W. Coates and H. G. Craighead, *Appl. Phys. Lett.*, 2003, **83**, 4836.
96. H. Dong, S. Prasad, V. Nyame and W. E. Jones, *Chem. Mater.*, 2004, **16**, 371.
97. H. Hou, Z. Jun, A. Reuning, A. Schaper, J. H. Wendorff and A. Greiner, *Macromolecules*, 2002, **35**, 2429.
98. M. Wang, N. Jing, C. B. Su, J. Kameoka, C.-K. Chou, M.-C. Hung and K.-A. Chang, *Appl. Phys. Lett.*, 2006, **88**, 033106.
99. H. Huang, R. M. Miura, W. P. Ireland and E. Puil, *SIAM J. Appl. Math.*, 2003, **63**, 1499.
100. X. Feng, A. K. Mairaj, D. W. Hewak and T. M. Monro, *J. Lightwave Tech.*, 2005, **23**, 2046.

CHAPTER 6

Single Molecule Analysis Using Single Nanopores

MIN JUN KIM[1], JOSEPH W. F. ROBERTSON[2], AND JOHN J. KASIANOWICZ[2]

[1] Department of Mechanical Engineering & Mechanics, Drexel University, Philadelphia, PA 19104, U.S.A.

[2] Semiconductor Electronics Division, EEEL, National Institute of Standards and Technology, Gaithersburg, MD 20899, U.S.A

6.1 INTRODUCTION

A significant obstacle for scientific progress in many biotechnology sectors, such as human health and medicine, is that the current cost of sequencing a mammalian-sized genome is between $10 million and $25 million.[1] In the previous two decades many efforts have therefore been focused on developing alternatives to the current DNA sequencing paradigm. Techniques based on single molecule sequencing may have the greatest potential for cost reduction. Unfortunately, to date, it has proven difficult to scale down DNA analysis to the single molecule level using conventional methods. Therefore researchers are exploring new approaches using single nanopores that can be adapted to active and smart structures. Such nanopores have tuneable sizes and controllable surface functionalities and have the potential to achieve low-cost and rapid sequencing of human genomes.

Macromolecules, such as DNA or RNA, are involved in the most important biological functions of the cell. As a result, characterizing and understanding their properties are crucial in advancing biomedical and biological research. Novel technologies to accurately and efficiently determine the sequence, structure and other properties of these molecules are always in high demand. The ability to determine DNA sequences has additional importance in identifying genetic mutations and polymorphisms. Recent advances in technology have made it possible to probe molecules one at a time. For example, atomic force microscopy (AFM) and laser tweezers have been used to obtain mechanical and kinetic properties of individual DNA, RNA and protein molecules, which have revealed novel information about their function, inaccessible using bulk measurements. However these techniques have not been successful in generating high resolution sequence information of macromolecules and their analytical throughput is generally low. Other single molecule analysis methods developed over the last decade use nanometer-sized pores in thin films and an electric field to induce the translocation of polymeric molecules through the nanopore, while changes in ionic current are monitored. The degree of current blockage as the molecule traverses the pore reflects the difference in the local structure along the polymer allowing sequence information to be obtained.

The rationale for using nanometer-sized pores to characterize biological macromolecules and polymer molecules is well known in the biological sciences. Several attempts have been made to adapt such nanopores for use in high-speed DNA sequencing.[2] Nanopore-based analysis methods are typically premised on the concept of passing a molecule, e.g. single-stranded DNA (ssDNA), through a nanoscopic opening while monitoring a signal.[2-6] Typically, the nanopore is designed to have a size that allows the ssDNA to pass in a sequential, single file order. As the ssDNA passes through the nanopore, differences in the chemical and physical properties of the nucleotides that compose the ssDNA are translated into characteristic electrical signals. The signal typically detected is a modulation of an ionic current, created by an applied voltage across the nanopore-bearing membrane or film, by the passage of the DNA through the nanopore. Because of structural differences between different nucleotides, each type of nucleotide interrupts the current in a different way, producing a type-specific modulation in the current as it translocates.[7,8]

The majority of the work performed in this area pertains to the use of a protein channel in a lipid bilayer.[3] It is known that proteinaceous nanopores, such as those formed by the toxin protein α-hemolysin secreted by the bacterium Staphylococcus aureus, possess a well-defined shape. Each pore is 1.5 nm at its narrowest point and consists of seven identical α-hemolysin molecules. The attractive features of protein nanopores include the well-established procedure for their synthesis.[2-6] The α-hemolysin protein nanopore is the archetype for rapid characterization and sequencing of nucleic acid molecules using high resolution analysis based on local light source and electrical read-out.[2,4,7] However, because it is not possible to tune the diameter of protein nanopores in both directions, investigating the structure, dynamics, and interactions of DNA/RNA molecules electrophoretically translocating through these ion channels has some limitations. It should also be noted that the α-hemolysin nanopore has a limiting aperture approximately 1.5 nm in diameter.[2] As a result, the pore is large enough to allow the passage of single stranded DNA, but too small to accommodate double stranded DNA.

An alternative to protein channels is the use of solid-state nanopores. These offer several advantages over phospholipid-embedded protein channels. The solid-state nanopores can be tuned in size with nanometer precision and also display an improved mechanical, chemical and electrical stability. However, the fabrication of these nanometer-sized pores on solid state materials represents a significant challenge, especially in the control and reproducibility of both the size and shape of the nanopores. Much research has focused on the fabrication of nanopores in solid-state thin films[9-14] and typically involves sophisticated instruments and complicated procedures. Furthermore, the mechanisms for solid-state nanopore formation are still not well understood, and the geometry and surface chemistry of such nanopores are not well characterized or controlled. The aims of this chapter are to outline some of the options that are available when choosing single nanopores for macromolecular characterization, and how they have been or can be manufactured using NEMS technologies.

6.2 FABRICATION OF SINGLE NANOPORES

6.2.1 Formation of α-Hemolysin Pores on Lipid Bilayers

Protein ion channels are nanometer-scale pores that span cell membranes. Depending on their size and function, they conduct the traffic of ions and/or macromolecules into and out of cells and organelles. The most well-known channels play key roles in nerve and muscle activity.[22] However, there is a class of pore-forming toxic proteins that are secreted by bacteria that spontaneously form relatively large nanopores in cell membranes. These include Staphylococcus aureus α-hemolysin[23] and Bacillus anthracis Protective Antigen.[24,25] The α-hemolysin channel has been used extensively as a model system to demonstrate proof-of-concept for a wide variety of analytical applications, including the detection, identification and quantification of ions,[26-29] proteins,[8] RNA and DNA polynucleotides.[2,7,17] A three-dimensional structure of the α-hemolysin nanopore obtained via x-ray crystallography is shown in Figure 6.1.

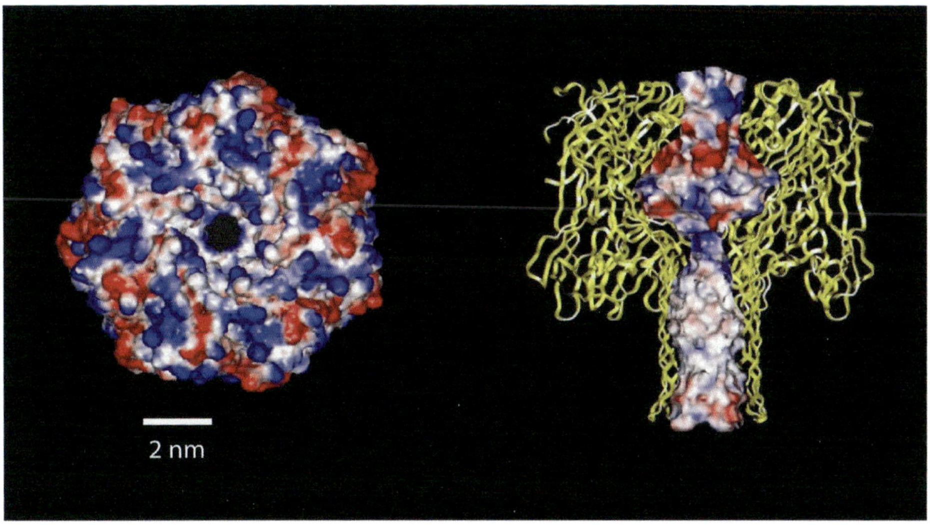

Figure 6.1 The 3-dimensional crystal structure of the α-hemolysin protein nanopore.[15] The narrow stem domain, which is shown in the cross-section of the side view (right), spans lipid membranes. The geometry of the nanopore has been verified using neutral and charged polymers.[2,8,16-21] The red, white and blue sections represent the negatively charged, neutral and positively charged regions on the channel respectively.

The high-impedance planar lipid bilayer matrix that the protein is reconstituted into is typically formed on a small hole in a 25 μm thick Teflon sheet that separates two aqueous electrolyte solutions (Figure 6.2). The membranes are made by either "painting" a solution of lipid (e.g. diphytanoyl phosphatidylcholine) in an organic solvent (e.g. *n*-decane) across the hole in the presence of the electrolyte solution,[32] or by the "solvent-free" method.[33] Planar lipid bilayer membranes have specific capacitances ~ 1 μF cm^{-2}. For membranes with diameters greater than 20 μm, the formation process can be monitored and quantified by applying a 10 mV (peak-peak), 100 Hz triangle wave voltage and measuring the resultant increase in the square-wave capacitive current ($I_c = C_m\, dV/dt$, where C_m is the membrane capacitance). The formation of much smaller membranes can be detected by applying a square wave voltage. A definitive test of whether a ~ 4 nm thick membrane has properly formed is to determine whether a self-assembling protein ion channel (e.g. gramicidin or α-hemolysin) can be easily reconstituted into the thin film. For high-fidelity nanopore-based analytical applications, the membrane conductance should be several orders of magnitude less than that of the protein ion channel.

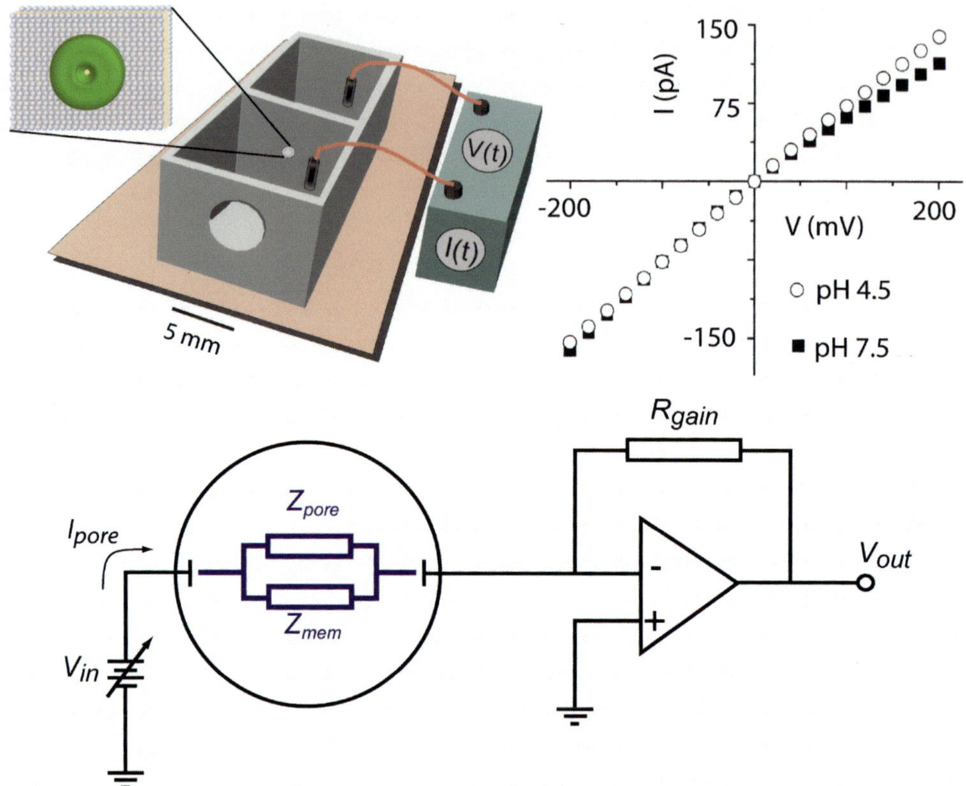

Figure 6.2 (Top left) Illustration of a typical system used to reconstitute protein ion channels into planar lipid bilayer membranes. A 4 nm thick membrane is formed over a small hole (~ 50 μm diameter) in a Teflon sheet that separates the two Teflon chamber halves that each hold ~ 0.2 mL to 2 mL of electrolyte solution. (Top right) Typical I-V relationships for a single α-hemolysin nanopore in the presence of 1 M KCl, pH 4.5 or pH 7.5. (Bottom) Schematic of a high impedance resistive feedback amplifier circuit used to convert the ionic current through the nanopore, I_{pore}, to a voltage (V_{out}). The membrane impedance, Z_{mem}, is measured before the addition of the pore-forming protein to one of the chamber halves. The nanopore ionic current, which is nearly ohmic, provides a crude estimate of the pore diameter (see text). The ionic current varies with pH because the fixed charges inside and near the pore bind positively-charged hydronium ions.[30,31]

A single α-hemolysin nanopore is formed by adding a small aliquot of the protein monomer, which is typically stored in 10 mM Tris, 1 mM EDTA, pH 7.5, to the bulk aqueous phase bathing one side of the membrane. The protein spontaneously binds to the lipid surface[23] and oligomerizes into a heptameric pore.[15,34] The conductance of a single α-hemolysin nanopore is ~ 1 nS in 1 M KCl, pH 7, V = 120 mV.[27] To ensure that only one channel forms, a dilute solution of protein is used. This causes a single pore to form in ~ 10 min with the chamber being flushed with an excess of protein-free solution after the formation of a single nanopore.

Planar lipid membranes are liquid crystalline and dynamic. They are not particularly robust in the sense that a static pressure gradient can rupture them. However, planar lipid membranes can easily persist for periods in excess of 24 hours. Conceivably, the membrane lifetime might be increased by forming the ultra-thin films on very small apertures[35] or by using lipids that can be polymerized.[36]

6.2.2 Formation of Solid-State Nanopores on Thin Films

6.2.2.1 Free Standing Thin Film Preparation

A prerequisite to making nanopores is the fabrication of solid-state free-standing thin films. There are many ways to make thin films. We have used the following methods to fabricate our free-standing thin films. Si_3N_4 films can be deposited using low pressure chemical vapour deposition (LPCVD) at a temperature of 825°C using ammonia and dichlorosilane gases.[37,38] Ammonia to dichlorosilane are introduced at a flow rate ratio of approximately 1:5. This results in a silicon-rich nitride film, with a tensile stress in the range of 50 to 150 MPa. This stress is low enough to allow the formation of free standing membranes. A 50 μm x 50 μm window is then fabricated on a ~ 350 μm thick silicon substrate wafer using photolithography and standard KOH wet-etching (Figure 6.3).[39,40]

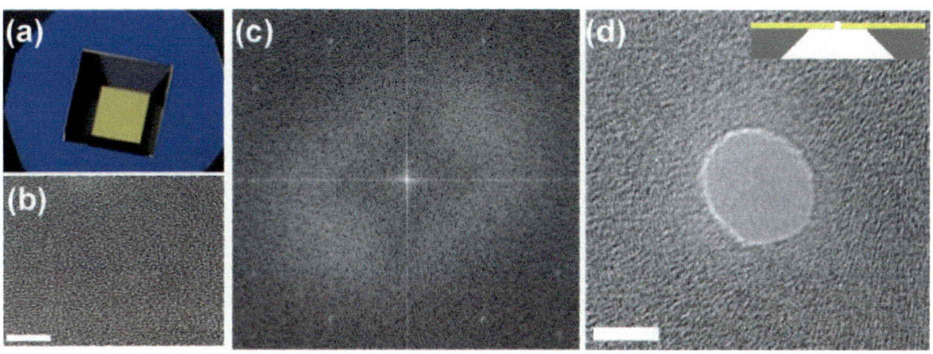

Figure 6.3 An image of silicon nitride film / chip (a), a TEM image of 50 nm thick silicon nitride membrane (b), a live fast Fourier transform (FFT) of silicon nitride membrane (c), and typical solid-state nanopore drilled by TEM (d). The inset in (d) is a schematic of the silicon nitride membrane supported by silicon. The scale bars are 5 nm in length. Adapted from M. J. Kim, *et al.*, Nanotechnology, 2007, 18, 205302 and reproduced with permission.

The mechanical and electrical properties of the free-standing film directly affect the efficiency and quality of the fabricated nanopores, as well as the performance of the nanopores when studying macromolecular translocations. We have performed a systematic evaluation of films that have different thicknesses, are made with different materials, and are deposited under different sets of parameters, to find general deposition rules and optimal conditions for thin film fabrication. The films can be characterized using ellipsometry, atomic force microscopy (AFM), and X-ray photoelectron spectroscopy (XPS) on the silicon substrates.

Intense e-beams (~10^8-10^9 e/nm^2) can be used to directly fabricate nanopores in the range of 4-8 nm diameter as previously reported.[41-44] For example, Figure 6.3 shows an 8 nm diameter nanopore. Depending on the thickness of the membrane (20-50 nm), pore formation times will vary between 5 and 60 seconds (with thicker membranes requiring longer times). Nanopore contraction is achieved by slightly defocusing the e-beam, effectively reducing the peak intensity to ~ 10^6 e/nm^2. This intensity is sufficient to fluidize the thin Si_3N_4 membrane, inducing rapid nanopore contraction. Figure 6.4 shows a sequence of TEM micrographs characterizing the dynamics of pore formation and contraction. The process enables the nanopore size to be tuned with an accuracy of ± 0.5 nm. With 1.5 to 2 nm diameter pores, high resolution analysis using an optical readout or ionic current blockade method can be used to study single-stranded DNA or RNA molecules. Pores with diameters in the range of 4 to 6 nm can be used to examine the structure, dynamics, and interactions of double-stranded DNA molecules.

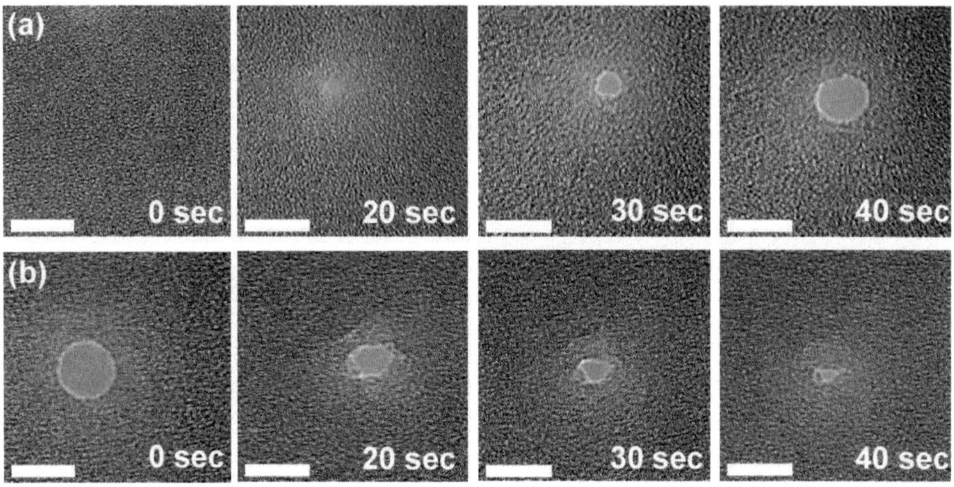

Figure 6.4 A sequence of TEM images displaying the dynamics of drilling (a) and contraction (b) in the 50 nm thick silicon nitride membrane. The scale bar is 5 nm in length. Adapted from M. J. Kim, *et al.*, Nanotechnology, 2007, 18, 205302 and reproduced with permission.

To further investigate the effect of surface tension induced contraction on nanopore kinetics, we fabricated and tested five different thicknesses of silicon nitride membrane (10, 20, 30, 40, and 50 nm). Nanopore contraction depends strongly on the thickness of the membrane, as well as on the initial size of the pore. These parameters were investigated by forming different sized nanopores using a high intensity e-beam, followed by intensity reduction to 10^6 e/nm^2. Figure 6.5 shows the process of nanopore contraction for two different initial sized nanopores (5 and 10 nm), as a function of the membrane thickness. These results clearly indicate that the contraction rate is greater for thinner membranes and that the contraction rate is constant. Interestingly, 5 nm pores made in 10 nm membranes, and 10 nm pores in 20 nm membranes cannot be contracted under any conditions.

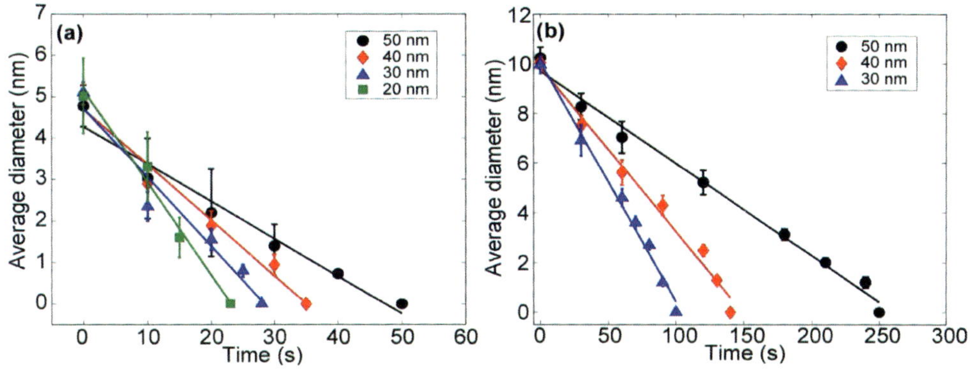

Figure 6.5 Nanopore contraction kinetics for a range of membrane thickness (20, 30, 40, and 50 nm) and two initial nanopore diameters. The initial size of nanopores are: (a) 5 nm, and (b) 10 nm. Nanopore contraction was not observed in the 10 and 20 nm thick membranes with initial sizes of 5 and 10 nm respectively. Adapted from M. J. Kim, *et al.*, Nanotechnology, 2007, 18, 205302 and reproduced with permission.

The linear dependence of the pore contraction rate on Si_3N_4 membrane thickness for the two nanopore sizes is shown in Figure 6.6. At a fixed e-beam intensity, the thinner the membrane the greater the contraction rate. The contraction rate depends linearly on membrane thickness with a slope of 0.0043 s^{-1} and 0.0030 s^{-1} for pores with initial diameters of 5 and 10 nm, respectively. In the most simplistic view, these rates should be approximately equal, since experiments were performed at constant e-beam intensity. The minor difference in the observed rates may be due to finite size effects. Previous studies suggest that contraction rates also depend on surface material properties.[41]

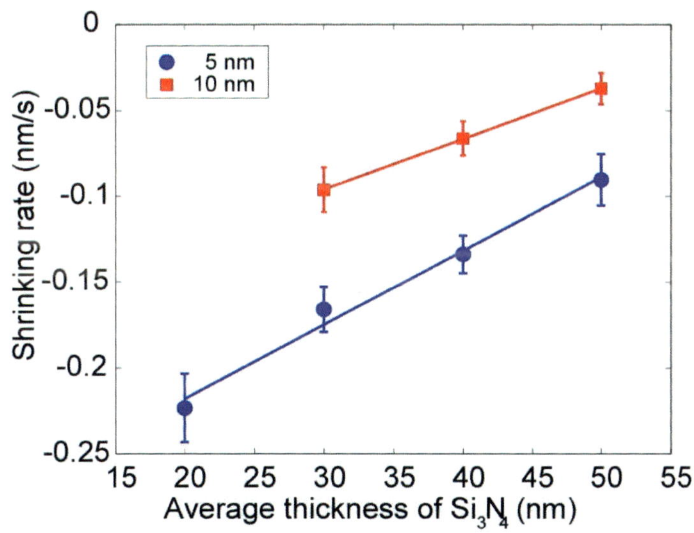

Figure 6.6 Variation of contraction rate as a function of the average thickness of the silicon nitride membrane (20, 30, 40, and 50 nm).

To explain nanopore contraction/expansion kinetics we introduce a simple model consisting of two processes: 1) nanopore expansion through atom sputtering, which is predominant at higher e-beam intensities, and 2) surface tension driven nanopore contraction, dominant at lower e-beam intensities. Experimental data show that the rate of nanopore contraction is linear in time, and that this rate is greater for thinner membranes. These results can be qualitatively accounted for by the following arguments. The sputtering rate (atoms sputtered per unit time) is a property of the material and a strong function of the e-beam intensity. If one assumes that the effective beam radius, φ, is larger than the initial nanopore radius, r, the resulting nanopore expansion rate is defined according to,

$$\frac{dr}{dt} = \tfrac{1}{2}\sigma\left(\frac{\phi^2 - r^2}{r}\right) : \phi > r \tag{6.1}$$

As observed experimentally for lower beam intensities, *dr/dt* is negative (i.e. the pore size contracts). This is due to the fact that σ is strongly suppressed, and surface tension effects become predominant. In this regime, the electron beam intensity fluidizes the thin Si_3N_4 membrane, and the nanopore can be approximated as an effective "hole" in a nearly two-dimensional fluid. Following recent analysis by Wu *et al.*[45] we approximate the Si_3N_4 membrane structure as a collection of infinitely thin two-dimensional disks. Furthermore these disks are assumed to contract with a uniform velocity and negligible interfacial friction between neighbouring disks. This simple model leads to a stationary nanopore contraction rate given by

$$\frac{dr}{dt} = \frac{\gamma}{4\eta}\left(-1 + \frac{r}{r_d}\right) \tag{6.2}$$

where γ is the surface tension of the (fluidized) Si_3N_4 membrane, η is its temperature-dependent viscosity and r_d is the radius of curvature along the cylindrical axis. Because the nanopore shape is nearly cylindrical we assume that in all cases, $r/r_d << 1$ and therefore the contraction rate is approximately $-\gamma/4\eta(T)$. The theoretically predicted constant contraction rates (independent of r) are supported by our results (which are in all cases approximately linear in time, Figure 6.5). Moreover, because the fluidized Si_3N_4 material exhibits a viscosity which is strongly temperature dependent, the steady-state temperature of the material determines the contraction rate. If the e-beam irradiated membrane temperature is determined by the balance of e-beam "absorption" and heat loss proportional to the cylindrical surface area of the nanopore (where $A = 2\pi rh$), then larger pores or thicker membranes will dissipate more heat and thus be cooler than smaller pores or thinner membranes. Therefore the membrane viscosity increases with A, and larger pores or thicker membranes should result in slower pore contraction kinetics, in agreement with the experimentally determined linear dependence shown in Figures 6.5 and 6.6.

Surface tension driven contraction also implies that there is a critical radius at which contraction does not occur, and for a membrane of thickness h contraction only occurs if $r < h/2$.[41,43] We have recently determined that for "hour glass" shaped nanopores the critical radius is determined by the effective membrane thickness, l shown in Figure 6.7.[44] For example, we have found that for a 50 nm thick membrane, l is found to be 30.4 $\pm$ 2.3 nm. Therefore based on the surface tension model the initial diameter of the pore can be no larger than ~ 16 nm (~ $l/2$) for contraction to occur.[44] This explains why we have not been able to observe nanopore contraction for 5 nm pores in 10 nm membranes, or 10 nm pores in 20 nm thick membranes (Figure 6.6).

6.2.2.2 Dimensional Structures of Solid-State Nanopore using TEM Tomography

To investigate the three-dimensional structure of solid state nanopores we utilized transmission electron microscopy (TEM). Specifically, samples were imaged at 40,000x magnification, with an underfocus of 2 – 4 μm, in a JEM2200FS TEM equipped with an Omega-type energy filter set to a slit width of 30 eV. A single axis tilt series of 46 images were recorded from –45° to +45° with a tilt increment of 2° using a Gatan Ultrascan 4k x 4k CCD camera in conjunction with SerialEM automated tilt series acquisition software.[46] The final pixel size of the image was 1.24 nm after an interpolation of the image size. Image processing of the tilt-series was carried out on a Linux workstation using the IMOD software suite.[47] Individual projection images of the tilt series were aligned with cross-correlation and the tomographic reconstruction was calculated by weighted back-projection.

When the high intensity electron beam impinges on the silicon nitride membrane, sputtering of Si and N atoms can occur on both membrane sides. This sputtering can result in pore formation.[41,42] The sputtering is angled on both sides of the membrane due to the intensity distribution around the central intense point, and consequently as sputtering continues an 'hour-glass' shape is created. As a result of this, the width of the pore is determined by the narrow section of this 'hour-glass'. This is illustrated in Figure 6.7, where this sputtering results in an angle, θ, which depends on the thickness of the membrane.

In the case of the 50 nm thick silicon nitride, the narrowest width of the nanopore is located 22 ± 6 nm above the bottom surface of the membrane. Further expansion of the pore, causes the narrow sharp edge of the 'hour-glass' to flatten. As is evident from Figure 6.7(e), a region of the membrane, l, becomes completely perpendicular to the membrane surface, and the sputtering angle, φ, on both sides of the membrane typically changes. The variables l and φ depend on the thickness of the membrane.

Reduction of the e-beam intensity, to ~ $10^6 e/nm^2$, results in a rapid contraction of the nanopore. At these lower energy beam intensities, atom sputtering is suppressed, and nanopore shrinking is driven by the surface tension effect of the fluidized atoms on the Si_3N_4 membrane.[41] To minimize surface energy, the atoms migrate to the flat region, l, of the nanopore resulting in a reduction in nanopore diameter.[41] This process is illustrated in Figure 6.4. Eventually this process may cause the pore to completely close. In the case of a 50 nm thick silicon nitride membrane, we find that is 132.7 ± 12.4° at the initial formation of the nanopore, while l and φ are 30.4 ± 2.3 nm and 25.3 ± 8.6° respectively after further expansion.

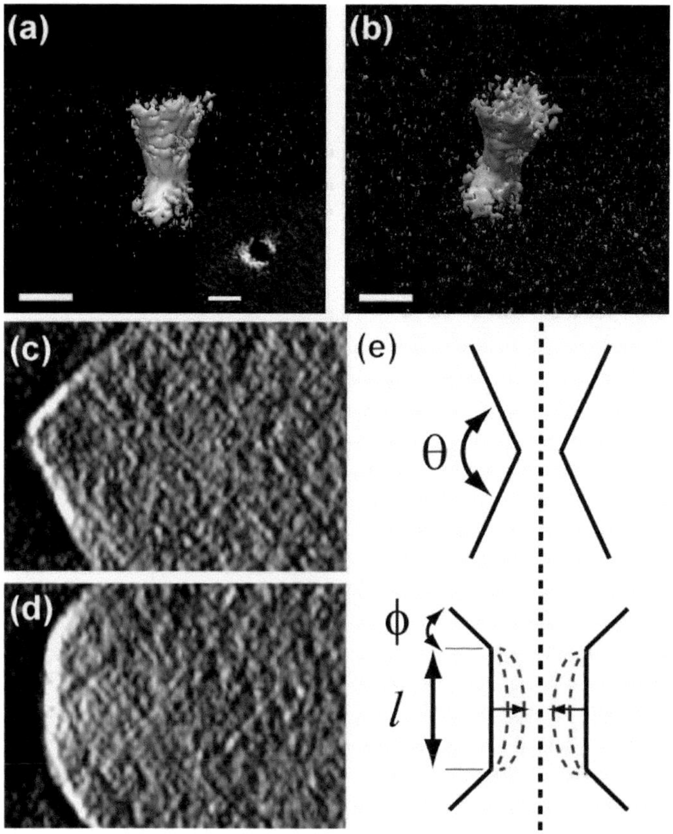

Figure 6.7 Typical three-dimensional structures of solid-state nanopores, as determined by TEM tomography.[11,44] (a) A cross-sectional image of a 8 nm diameter nanopore (inset: top view), (b) a tilted view, 25° tilt in z-y and z-x planes, (c) a bisectional view of initial formation of solid-state nanopore, (d) that of a further expanded nanopore by high intensity of electron beam, and (e) a schematic of the cross-sectional view of the initial nanopore and of a nanopore that has been expanded and then shrunk (red dotted lines). The scale bars are 10 nm. REFERENCE

6.2.3 Experimental Setup for Ionic Current Blockade Measurements on Nanopores

6.2.3.1 α-Hemolysin Nanopores

As described previously, planar lipid bilayer membranes can be formed on a small orifice in a thin Teflon sheet that divides two halves of a Teflon chamber (Figure 6.2). The ionic current through a single α-hemolysin nanopore is measured by applying a transmembrane electric field ($E = -dV/dx$) across the membrane via two Ag-AgCl electrodes and conversion of the signal to a voltage with a high-impedance amplifier.[27] The signal is digitized using a 16-bit A/D converter and analyzed with in-house software. The magnitude of the applied potential is typically less than 250 mV to minimize the possibility of rupturing the membrane.[48]

6.2.3.2 Solid-State Nanopores

Inspection of Figure 6.8 shows that the flow cell consists of two chambers; an upper chamber (blue) and lower chamber (green), both of which are made entirely out of polytetrafluoroethylene (PTFE) due to its low noise rating and chemically inert properties. The flow cell is prepared by cleaning both the chambers and the electrode supporting ring (yellow) with Piranha solution for 15 minutes and thoroughly rinsing with filtered Millipore water, methanol, and drying under a N_2 stream. After cleaning, a 5 × 5 mm^2 nanopore chip (red) is adhered to the upper chamber using a fast curing silicone elastomer epoxy. A 0.25 mm diameter PTFE coated silver electrode with an exposed tip is then placed and sealed using silicon elastomer epoxy with its tip 1.5 mm inside the lower chamber. After curing the lower chamber is filled with a 0.002 M KCl solution. The upper chamber is placed on the lower chamber and fixed in place using PTFE screws.

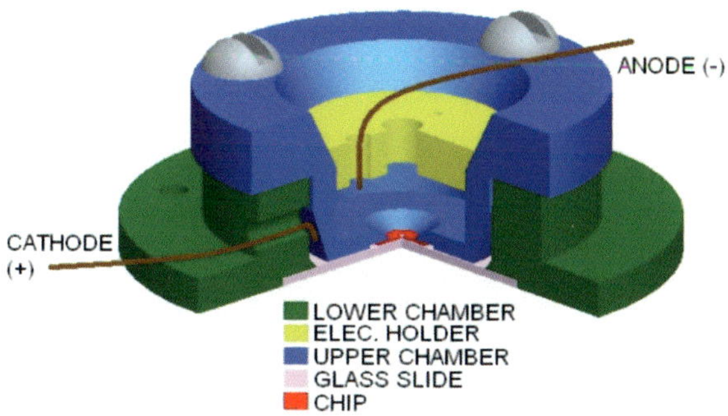

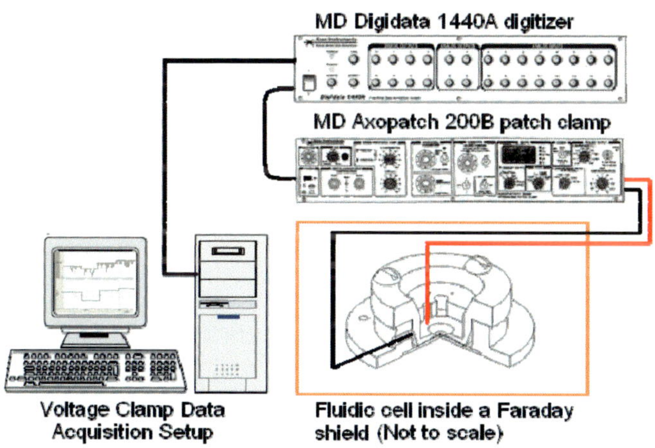

Figure 6.8 (Top) Rendering of experimental flow cell with a 90° radial cut-away for cross-sectional view. Upper chamber (blue), Electrode supporting ring (yellow), Lower chamber (green), Micropore chip (red), and glass slide (pink). (Bottom) Schematic of entire experimental set-up.

Another short PTFE coated silver electrode with an exposed tip is threaded through the electrode supporting ring exposing its tip 1.5 mm beneath the ring's bottom. The upper chamber is then filled with a 0.002 M KCl solution and the support ring is placed into its seat. This process defines the distance between the two electrodes. The completed flow cell is then screwed in place within a 5 mm thick copper Faraday shield to reduce any electromagnetic field induced noise on the detected electrical signal.

Voltage clamping of the micropore is performed using a Molecular Devices Axopatch 200B patch clamp in 'whole cell' mode allowing the direct observation of fluctuations in ionic current through the micropore. Data are acquired by the patch clamp and digitized using a Molecular Devices Digidata 1440A digitizer and analyzed using pClamp 10.1 software (Figure 6.8). Signals are sampled at 200 kHz and filtered at 10 kHz using a low pass Bessel filter. To estimate the nanopore conductance, an I-V curve is obtained by measuring the current response to incremental increases in command voltage.

6.3 ANALYSIS OF NUCLEIC ACIDS USING NANOPORES

As a polymer, DNA consists of repeated units, specifically four nucleotides of different sizes, shapes and chemical composition. These form a backbone together with each sequential nucleotide's phosphates and sugars. DNA can exist in a double stranded helical state or a single stranded state in which a complementary strand is removed. While models have been developed to understand the mechanics of DNA on a macromolecular level, little is known about the mechanics of DNA on the nanoscale where even the local sequence of nucleotides will affect the molecule's mechanical characteristics in a narrow area. Because of these unknowns analyzing the sequence of ssDNA molecules has proven difficult. However many international research programs have been working towards this goal. We describe below experimental set-ups that have been and are currently used in ionic current blockade analysis of macromolecules for α-hemolysin and solid-state nanopores.

6.3.1 Characterization of Single Nanopores

6.3.1.1 α-Hemolysin Nanopores

The three-dimensional x-ray crystal structure of α-hemolysin nanopores removed from lipid membranes has been determined to 0.19 nm (Figure 6.1).[15] The axial length is 10.5 nm, and the diameter varies from ~ 4.0 nm to 1.6 nm inside the pore. The results of a variety of electrophysiological measurements, described in this section, are consistent with this structure.

A crude estimate for the diameter, *d*, of a nanopore can be obtained by assuming the pore is a uniform circular cylinder with cross-sectional area $A = \pi d^2/4$ and length *l*. The resistance of this idealized nanopore is $R = l/A\sigma_{pore}$, where σ_{pore} is the conductivity of the electrolyte solution inside the nanopore. The single channel conductance, $g = 1/R$, of the α-hemolysin nanopore is 1 nS in 1M KCl.[2] If we assume that the conductivity of the solution in the pore is identical to that in the bulk aqueous phase ($\sigma_{bulk} \sim 100$ mS cm), then $d = 2(lg/\pi\sigma_{bulk})^{1/2} \sim 1.12$ nm, or slightly less than smallest diameter in the nanopore crystal-structure model.

At the nanometer-length scale, factors other than geometry can affect ion transport. For example, the crystal structure of the α-hemolysin nanopore locates rings of ionizable amino acid side chains (histidines, glutamic acids, aspartic acids) inside or near the pore. These are likely to affect the local electrostatic potential along the pore axis and therefore the local concentration of pore-permeant ions. The value of the charges on these side chains will be altered by changes in the pH of the solution. Indeed, a decrease in the pH from 7.2 to 4.5 causes the conductance for positive applied potentials to increase by about 10% in the presence of 1M NaCl (Figure 6.2).[27] Also, chemical modification of site-directed mutants of α-hemolysin demonstrates experimentally how the location of charges inside the pore can alter the single channel conductance and ion selectivity of the nanopore.[31]

The diameter of the α-hemolysin nanopore has also been estimated using differently sized non-electrolyte polymers of polyethylene glycol (PEGs).[16,49-52] The principle of this method is simple. PEGs added to electrolyte solutions decrease the bulk conductivity. Therefore, PEG molecules that are small enough to enter the pore should decrease the pore conductance. Those that are too large to enter the pore will not. Figure 6.9 shows qualitatively the results of such an experiment with the α-hemolysin nanopore. The PEG molecular mass cut-off is ~ 2,200 g mol^{-1}. At this size PEG has a hydrodynamic diameter of ~2.5 nm,[16,53] which is greater than the smallest diameter, but less than the largest diameter of the α-hemolysin nanopore crystal structure.

The crystal structure of the α-hemolysin nanopore (Figure 6.1) suggests that the smallest constriction inside the pore is located closer to the stem domain entrance than to the cap domain. This has been confirmed using different length "molecular rulers" comprised of a polynucleotide attached to a large macromolecule, as illustrated in Figure 6.10.[21] Specifically, different length polynucleotides biotinylated on one end were attached to streptavidin and added to the electrolyte solution bathing the cap or stem sides of the α-hemolysin nanopore. Shorter polynucleotides attached to streptavidin were driven into the pore temporarily by an applied potential. Complexes with polynucleotides longer than a specific length, which depended on the side to which they were added, were driven into the nanopore and stayed there as long as the potential was applied. Unlike the shorter polynucleotides, the longer polynucleotides were most likely forced past the narrowest constriction inside the nanopore. Once past that location, polynucleotides will be lodged there until the applied field is turned off or reversed.

Over a decade ago, we demonstrated that single-stranded RNA and DNA (ssRNA and ssDNA) can be driven electrophoretically through a single α-hemolysin nanopore.[2] Polynucleotides at concentrations of 0.1 nM or greater cause conductance blockades that are well-defined in both amplitude and lifetime.

As will be discussed subsequently, conductance blockades caused by some homopolymeric ssDNAs are relatively simple. Specifically, the blockade amplitudes are predominately described by a single mean value,[2,8] and the residence time distributions for polymers in the pore are Gaussian in nature.[2] This is generally not the case for polythymines. Specifically, poly[T] strands which are monodisperse in length (e.g. 100-nucleotides long) have been shown to cause blockades with multiple levels that are stationary. This behaviour is shown in Figure 6.11 (top).[8,18,20]

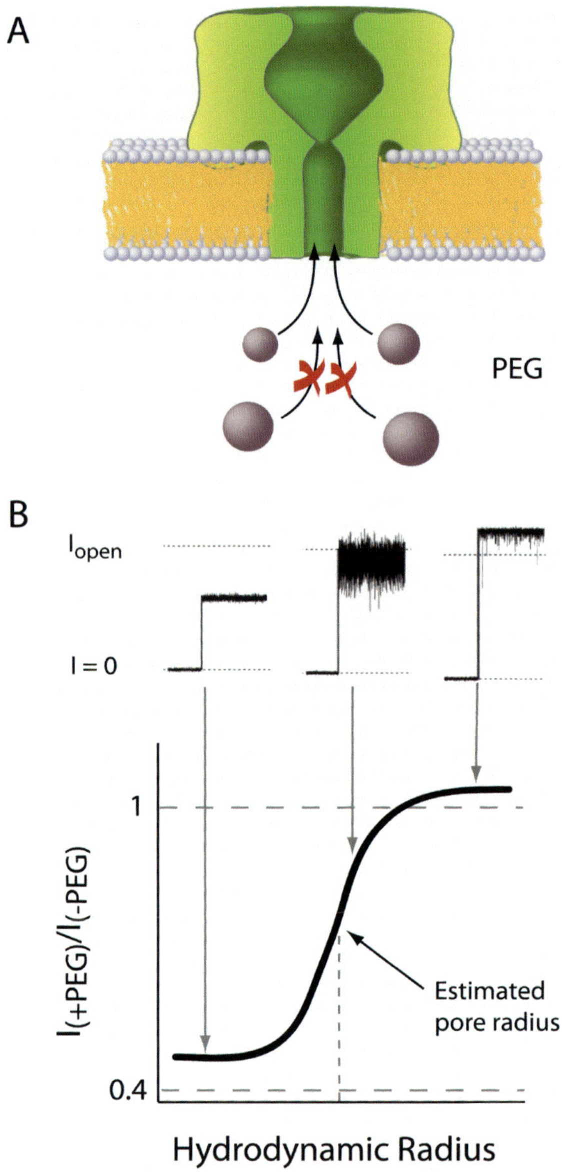

Figure 6.9 Water-soluble, nonelectrolyte polymers (e.g., poly(ethylene glycols), PEGs) are commonly used to estimate the limiting aperture diameter of protein nanopores. (Top) Only PEGs that are smaller than the pore diameter will partition into it. (Bottom) When inside the nanopore, PEG molecules reduce the nanopore ionic current.[50,51] The diameter of the pore is estimated from the hydrodynamic radius of the largest-size PEG molecules[53] that enter the pore. PEGs cause strong current fluctuations because they bind to the interior of the α-hemolysin nanopore.[16,50,54]

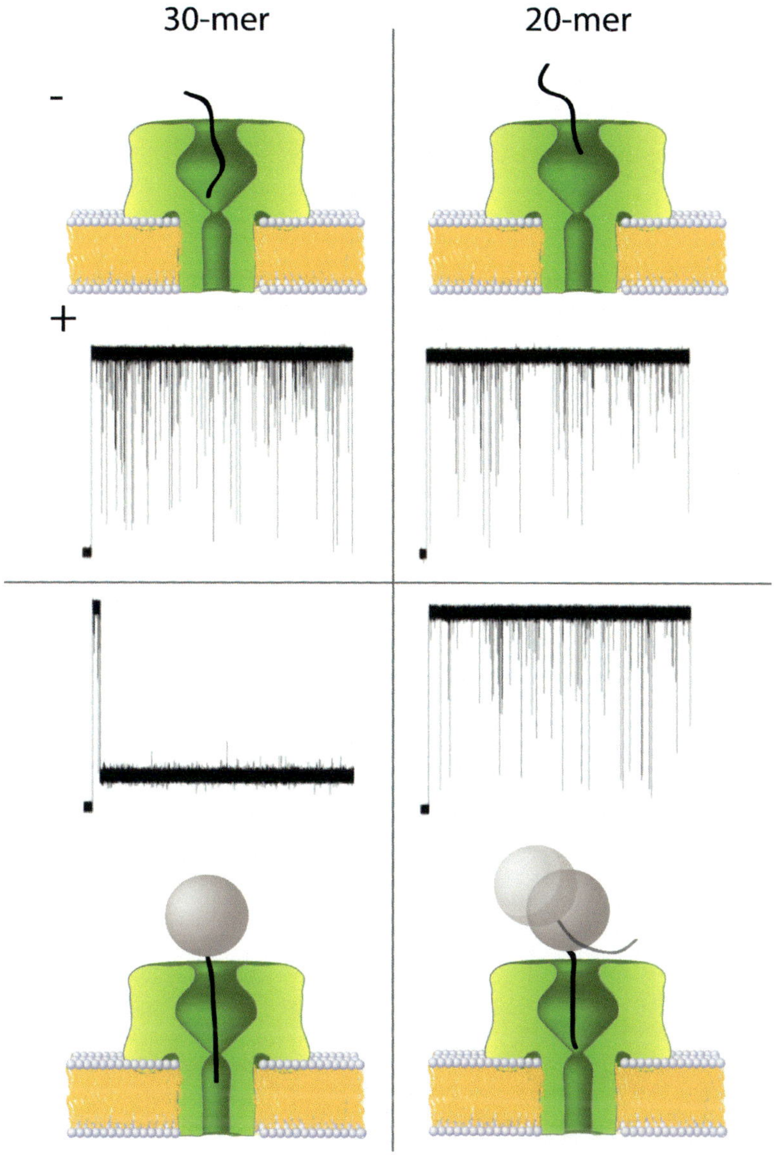

Figure 6.10 The use of DNA "molecular rulers" to locate the smallest aperture inside the α-hemolysin nanopore. (A) Two different length poly[dC] homopolynucleotides, biotinylated on the 5'-end, were added to the electrolyte solution bathing one side of the nanopore and caused transient blockades in the ionic current. The subsequent addition of avidin, which binds strongly to the biotin, caused 30-nucleotide long polynucleotides to occlude the pore indefinitely when the applied potential was constant. However, the shorter strand (i.e., 20-nucleotide long) did not. Qualitatively similar results were obtained when the polymeric rulers were added to the opposite side. The data suggest there is a tight constriction deep inside the nanopore over which the applied potential drops significantly.[18,21]

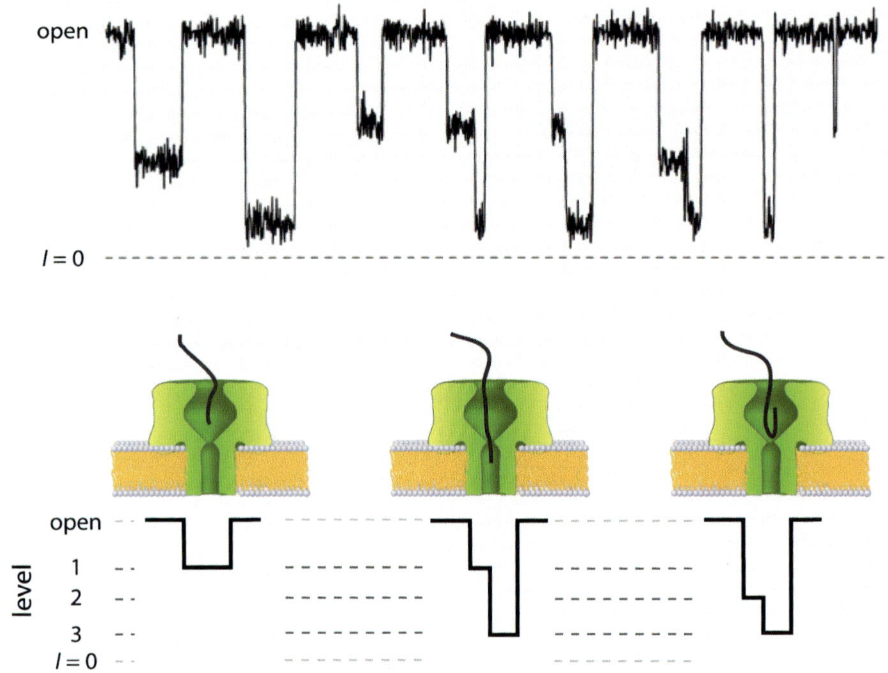

Figure 6.11 (Top) Ionic current blockades of the α-hemolysin nanopore caused by 100 nucleotide-long poly[dT] molecules. The polymer was added to one side and the applied potential was V = -120 mV. The patterns of the current blockades, which illustrate the states of the bimolecular interactions between the polymer and various parts of the pore interior, are consistent across large ensembles and can be classified using advanced statistical signal processing.[18-20,54,55] (Bottom) The blockade morphologies correlate with the geometry of the channel.[18,20,21]

Curiously, some of the poly[T]-induced blockades lasted approximately 1 s. However, the duration of most blockades was ≤ 2 ms. The blockade amplitudes of these events were modeled using a five-kernel Gaussian Mixture Model (GMM), and are shown in the schematic representation of the ionic current in Figure 6.11 (dashed lines, bottom). As is illustrated in the cartoons, two of the three distinct poly[T]-induced blockade amplitudes are probably caused by two different amounts of polymer mass in the cap domain vestibule of the α-hemolysin nanopore. The third amplitude is caused by the polymer threading into the narrowest aperture.[21]

Interestingly, for polymer added to the cap domain side of the nanopore, the poly[T]-induced pore conductance blockade structure typically evolves from either state 1 to state 3, or state 2 to state 3. When poly[T] was added to the stem entrance side of the pore, the multi-state blockades evolved from state 3 to state 1 or state 3 to state 2 (not shown).[21] This is consistent with the polymer entering the cis-side vestibule, being dislodged in the short events, and progressively exploring the narrower apertures further down the channel in events with increasing lifetimes. These results are also consistent with the channel structure, since the cap-domain vestibule is larger than the pore aperture closer to the stem-side entrance. For a given length of polymer, deeper conductance blockades will be observed in a narrower segment of the nanopore.

6.3.1.2 Solid-State Nanopores

We have characterized the ionic conductivity of our nanopores by performing a set of I-V measurements at different ionic strengths. In all cases, (3 nm to 15 nm) the nanopores exhibit Ohmic behaviour with applied potentials between ± 0.5 V. The slope of the I-V relationships is used to determine the pore conductance, G. Based on TEM tomography information (Figure 6.7), an equation for the ionic conductivity through a truncated double-cone structure (excluding surface-induced flow due to the shielding counterions) can be derived, i.e.

$$G = \frac{\pi d^2}{4}\sigma\left(\frac{\delta \tan\alpha + 1}{h + h_{eff.}\delta \tan\alpha}\right) \tag{6.3}$$

where $\sigma = (\mu_K + \mu_{Cl})n_{KCl}e$ is the specific conductance at a number density n_{KCl}, equal to 15.04 (ΩM^{-1}) and 3.01 (ΩM^{-1}) for 1 M and 0.2 M KCl, respectively, (μ_K and μ_{Cl} are the electrophoretic mobilities of K^+ and Cl^-, respectively, and e is the elementary charge unit), $\delta = (h - h_{eff.})/d$, $h_{eff.}$ is the width of the cylindrical region in the nanopore, and α is the cone half angle. The dependence of G on the pore diameter is shown in Figure 6.12. The inset depicts an idealized model for the nanopore structure. Using known values for σ and $h = 50$ nm, the data can be fitted using Equation 6.3 with two free variables (h_{eff} and α). From these fits (solid lines) we obtain $h_{eff} = 17 \pm 1$ nm and 18 ± 1 nm for the 1 M and 0.2 M data, respectively, and $\alpha = 30° \pm 2°$ for both cases.

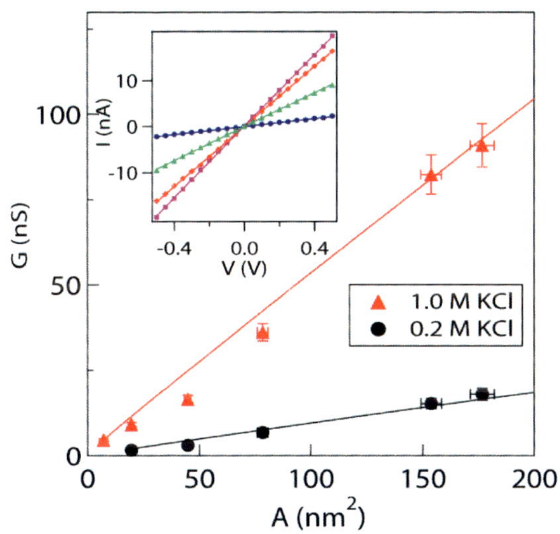

Figure 6.12 Ionic conductance (G) of solid state nanopores of various cross-sectional areas (A) at 1.0 M (triangles) and 0.2 M (circles) KCl solution (+10 mM TRIS buffer, pH 8.5).[11] The lines represent fits to Equation 1 assuming a cylindrical pore geometry, revealing a pore length L = 30 ± 2 nm. Inset: I-V curves obtained at 1.0 M KCl for nanopores with average diameters (± 5%) of 12 nm (squares), 10 nm (diamonds), 7 nm (triangles) and 3 nm (circles).

The results obtained from Figure 6.12 are in a good agreement with the three-dimensional reconstructed image obtained by TEM tomography (Figure 6.7). In particular, we note that $h_{eff} / h \sim 2.9$, a value close to our estimation from the TEM image. Figure 6.12 shows a linear dependence of the pore diameter on the square root of the ionic conductivity, indicating the effect of reducing the size of the pore on the resistance to ion transfer. The ratio of the slopes for the two lines obtained for different concentrations is 2.21, compared to 2.23 expected for a five-fold increase in the electrolyte concentration.

6.3.2 Analysis of Single Molecules Translocating Through Single Nanopores

6.3.2.1 α-Hemolysin Nanopores

Unlike many protein ion channels that spontaneously switch or "gate" between different conductance states, the α-hemolysin nanopore can remain fully open for long periods of time.[27] This feature, the size of the limiting aperture diameter inside its pore (1.6 nm), and the relative paucity of fixed charges inside the pore[15,30,31] makes the α-hemolysin nanopore ideal for the identification and quantification of analytes[27-29] and for studying the physical properties of polymers (e.g. ssRNA, ssDNA, proteins, PEGs).[2,8,16-18,54]

In the absence of polynucleotides, the ionic current that flows in response to a fixed applied potential is large and stable. After adding polynucleotides of uniform length (either ssRNA[2] or ssDNA[8,17]) to the electrolyte solution either side of the membrane, well-defined reductions in the current occur. This effect is clearly seen in Figures 6.11 and 6.13. For homo-polynucleotides of ssRNA comprised of the same base (e.g. poly[U]), the mean residence time of the polymer in the pore is observed to increase in proportion to the polymer length.[2] Curiously, each monodisperse-length of poly[U] caused current blockades having one of three characteristic residence times. The shortest mean residence time is most likely caused by the polymer leaving the same entrance of the nanopore that it enters. The two longer residence time distributions are thought to be caused by the transport of polymers entering the pore with either their 3'- or 5'- ends.[2] Since the two longer residence times of the polymer in the pore are observed to increase linearly with the number of bases in the polymer, the ssRNA molecules most likely thread completely through the pore. PCR analysis has confirmed that ssDNA, but not dsDNA can be transported through the channel.

A transmembrane concentration gradient of polynucleotides alone is insufficient to initiate polynucleotide transport through the nanopore. A minimum voltage gradient (~ 60 mV) is required to drive the polymers over a barrier inside the α-hemolysin nanopore.[17,56] Not surprisingly, this barrier is located at the narrowest constriction inside the pore.[18,21] Because polynucleotides are driven through the α-hemolysin nanopore as a relatively straight rod and cause easily observable ionic current blockades, it has not surprisingly been suggested that it might be possible to sequence DNA by measuring the current time series of the polymer-induced conductance blockades, assuming that each base gives rise to characteristic blockade amplitudes.[2] If possible, sequencing long DNA strands could be performed rapidly since the residence time for any given base in the pore is < 10 μs.

Homopolymers of ssRNA poly[C] and poly[A] have been shown to cause distinctly different blockade distributions. Specifically, poly[C] RNA blockades are on average deeper in amplitude and shorter lived than those induced by poly[A] RNA.[7] In addition, a di-block copolymer of poly[C]:poly[A] primarily generates two-step conductance blockades that correspond to the poly[C] segment entering the pore first, followed by the poly[A] section. One might be tempted to suggest that this result provides evidence for base-by-base polynucleotide sequencing. However, although cytosine is smaller than adenosine, poly[C] blocks the pore conductance more than poly[A]. In this case, the α-hemolysin nanopore did not discern the difference in volume occupied by a given base on a polymer. Instead, the pore may have been sensitive to gross structural features. For example, if poly[C] (poly[A]) forms helices with a mean diameter that is smaller (larger) than the pore's smallest aperture,[57] then poly[C] could be driven through the pore unfettered, but poly[A] would have to unwind before it threads through the channel. If that were the case, then poly[C] should perturb nanopore conductance more than poly[A] since the former would have more mass inside the nanopore per unit pore length. Moreover, poly[C] would be driven through the pore at a faster rate because the charge per unit length would be greater in a helical structure than in an extended polymer.

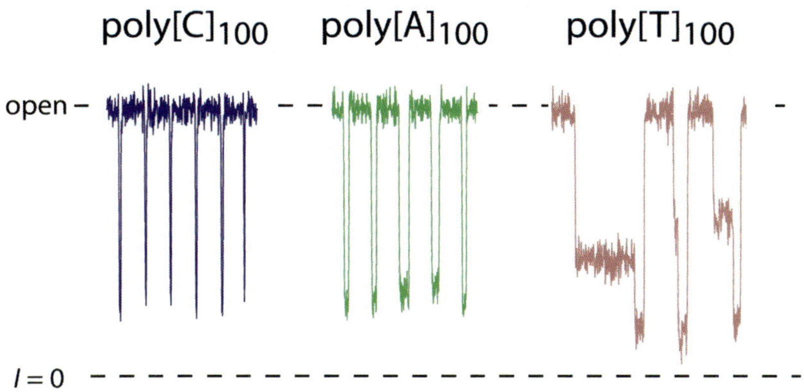

Figure 6.13 Different ssDNA polymers cause distinct and distinguishable classes of blockades in the single α-hemolysin nanopore ionic current. The different interactions between identical length homopolynucleotides of poly[C], poly[A], and poly[T], each 100 bases long, is evident as judged from the blockade amplitudes, residence times for the polymers in the nanopore and the patterns of current blockades.

There may be alternative ways to sequence DNA other than by simply reading the current-time series in a serial manner. When the rate at which polymers translocate a pore is controlled only by the applied potential and the stochastic interactions between the polymer and the nanopore (which is typical for most research reported in this field to date), the blockades caused by a given polynucleotide class will not be identical (Figure 6.13). However, despite the variations from blockade to blockade, there are clear features characteristic of the polymer type. For example, blockades caused by 100-nucleotide long ssDNA poly[C], poly[A] and poly[T] vary from event to event for each given polymer, but there is no mistaking the fact that blockades caused by one polymer are quite different to those caused by another. The blockade patterns caused by poly[T] are relatively simple for polymers with residence times < 2 ms (Figure 6.11). Specifically, the open, closed and polymer-induced blockades are described by a 5 component GMM. Some poly[T] blockades persist for ~ 2 s. These longer blockades require a GMM with ~ 38 amplitudes! This degree of complexity may seem to be a disadvantage. However, the more components that are needed to classify signals caused by a given polymer may make discrimination of different polymers possible.[18,19,58]

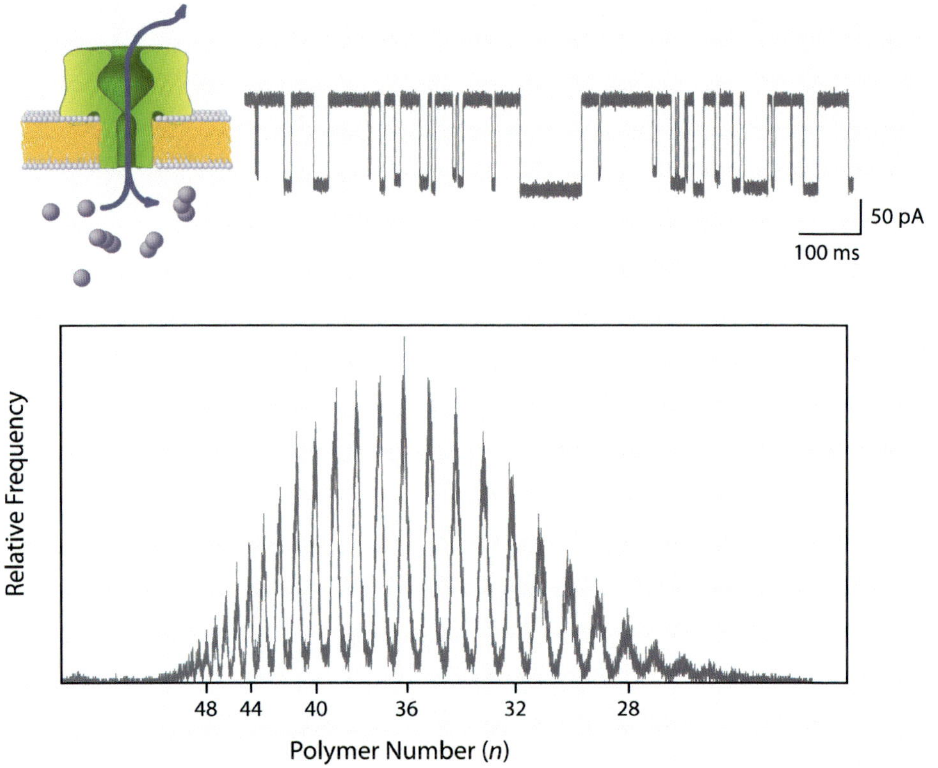

Figure 6.14 A single α-hemolysin nanopore was used to determine the mass spectrum of a polydisperse polymer sample. PEG molecules added to the aqueous phase bathing one side of the membrane partition into the pore (Top left) and reduce the pore conductance (Top right). A histogram of the mean values for each conductance blockade (Bottom) correlates with a MALDI-TOF spectrum of the same sample (not shown) and clearly can distinguish between PEG molecules that differ by a single ethylene glycol. Monodisperse PEGs are used to calibrate the conductance-based mass spectrum (not shown) and demonstrate that larger polymers block the current more than smaller ones.

The α-hemolysin nanopore can detect and quantify proteins.[8,18] Recently, it has been used to physically characterize other solution phase macromolecules (e.g. synthetic polymers such as PEG). The detection scheme for polymer analysis is identical to that for nucleic acids, as is illustrated in Figure 6.14 (top left). A polymer sample is added to one side of the nanopore membrane, and the current fluctuations caused by individual polymers interacting with the pore are observed (top right). In contrast to many polynucleotide-induced conductance blockades, each blockade caused by a single PEG molecule in the nanopore is characterized by a single Gaussian. A histogram of the mean current values for each of the polymer-induced blockades produces a conductance-based analogue of a mass spectrogram of the sample (bottom). Calibration of this spectrum was achieved using a highly-purified sample of PEG (not shown). The physical basis for separation lies in the sensitivity to the subtly different solvent exclusion volume of differently sized polymers.

Because measurement resolution relies on accurately predicting the mean current, extending the residence time of the polymer in the nanopore is essential to achieving high resolution. As a first order approximation, the interaction between pore and polymer can be treated as a simple reversible adsorption reaction, which is dependent upon the chemistries of both the polymer and the pore. The equilibrium constant of these interactions can be adjusted by changing solution conditions to optimize the residence time of the polymer of interest. In the case of PEG, increasing the electrolyte concentration increases the residence time due to increased screening of the fixed charge within the α-hemolysin nanopore.[16,51,54,59] Further enhancement of the signal can be achieved by selectively observing only long events (for example those longer than ~ 1 ms).

6.3.2.2 Solid-State Nanopores

To demonstrate the functionality of solid-state nanopores, we have performed DNA translocation experiments, similar to those reported for α-hemolysin[4] and other solid state pores.[12,41,60] Each nanopore chip was mounted in a custom-built cell, similar to that described in Figure 6.8, to enable low-noise electrical measurements. These cells form miniature "*cis*" and "*trans*" fluid chambers, accessible by fluid lines.

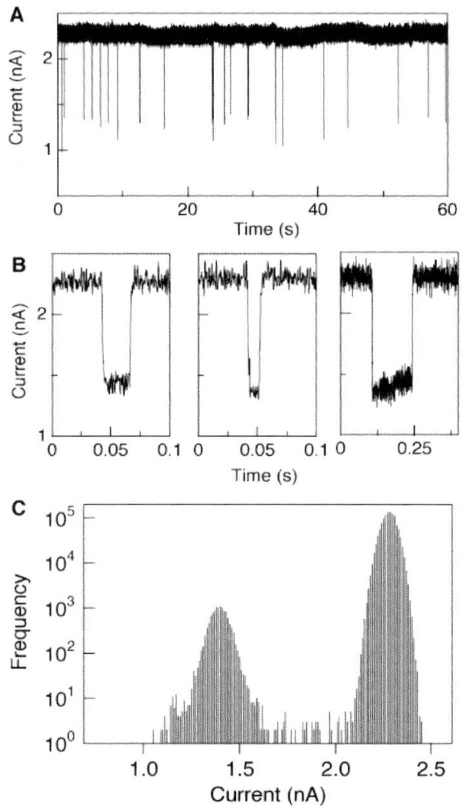

Figure 6.15 (A) Typical ionic current trace of a 4 nm nanopore at 1 M KCl after the addition of 10 nM double stranded DNA (150 basepairs), measured using a bias of 400 mV.[11] Discrete drops in the ion current are clearly observed, corresponding to pore blockades due to the translocation of the DNA. (B) An expansion of three typical translocation events. The abrupt changes in the current upon DNA entry and exit from the nanopore occur in less than 50 μs. (C) All-point histogram of the ion current displays two prominent peaks: at 1.4 ± 0.1 nA (blocked level) and at 2.3 ± 0.1 nA (open pore).

The two chambers were fitted with ports for Ag/AgCl electrodes. The signals were filtered at 20 kHz using a Butterworth low-pass transfer function and digitized at 100 kHz (12 bit). Figure 6.15 displays ion current blockades produced by applying 0.01 μM of 150 bp double stranded DNA (PAGE purified) to the negative chamber of a chip (the "*cis*" chamber) containing a single 4 nm nanopore. Upon the addition of DNA, the ionic current developed sharp blockades spikes, similar to those observed for polynucleotide transport through the α-hemolysin nanopore.[4] A typical set of ion current blockades is shown in Figure 6.15(B). Figure 6.15(C) displays a typical all-point histogram of the ion current. Two peaks are readily visible; a high peak and a low peak corresponding to the open pore current (2.3 nA) and the blocked pore current (1.4 nA) respectively. This yields a normalized blocked current value of 0.61 nA, in agreement with the expected blockage based on the DNA and nanopore dimensions. Similar DNA translocation experiments were performed for other nanopore sizes in the range 3 - 10 nm. We note however that above ~ 4 nm dsDNA produces additional peaks in the current histogram, which can be interpreted as DNA entries in partially folded configurations.[12,61]

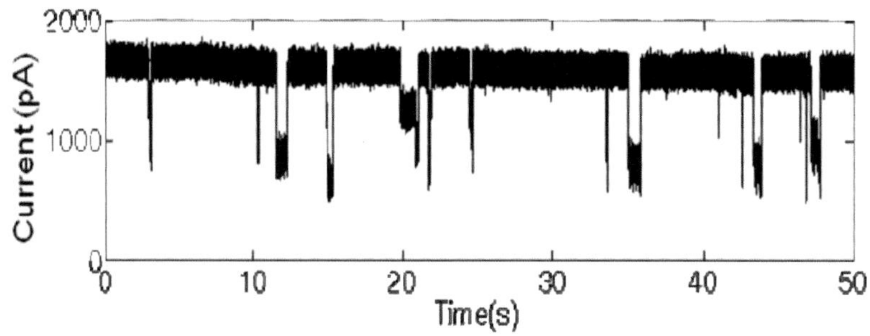

Figure 6.16 Current trace of a of a 5 nm nanopore at 1M KCl concentration after the addition of 10μl of PCR product solution containing dsDNA (~100bp) and PCR enzymes, measured using a bias of 400mV.

Additionally, in a recent unpublished experiment we added PCR product solution containing dsDNA (~100bp) and PCR enzymes to a flow cell fitted with a 5 nm pore. A 400 mV command voltage was used to translocate the PCR solutes, resulting in the data plot of Figure 6.16. This demonstrates a wide distribution of current blockade dwell times and shapes reflecting the variety of solutes in the solution. This proof-of-concept experiment shows the nanopore's ability to not only detect a separated analyte but also a mixture of differently sized, shaped and chemically composed macromolecules. In future work we hope to distinguish the sub-characteristic dimension morphology of different macromolecules in sample solutions based on their "signature" current blockade, thereby detecting the entire contents of the solution.

6.4 CONCLUSIONS

The technologies for sequencing, characterizing and utilizing nucleic acids have been dramatically advanced over the last decade, however the demand for rapid, low cost and accurate methods to determine various properties of nucleic acids is still pressing. In this chapter we have highlighted the unique and complementary capabilities of nanopore technology in this respect by measuring the sequence, physicochemical and structural properties of single molecules.

The α-hemolysin channel has proven to be an important model system for establishing the validity of a wide range of nanopore-based analytical applications. These include the detection, identification and quantification of ions, proteins and nucleic acids. Although biological nanopores have some potential drawbacks for practical analytical applications, some (notably the α-hemolysin channel) have extremely useful features. For example, the pores self-assemble into highly reproducible structures, a feat that top-down materials science has yet to achieve.

As discussed, much effort has been devoted to making single digit nanometer size pores on solid-state thin films. Nevertheless, these techniques are in their infancy. The control over the size and shape of individual nanopores is not yet robust, and the instrumentation used is specialized. Indeed the use of solid-state nanopores poses experimental challenges, such as inherent pore-to-pore inconsistency and unfavourable surface properties (which can lead to inferior pore reproducibility), pore clogging and larger electrical noise. In overcoming the challenges associated with forming nanoscale pores of controllable size and characteristics suitable for probing DNA and other biological molecules, a novel scanning TEM fabrication approach has been shown to provide for uniform sized nanopores and the ability to automate with no compromise in accuracy.[11] A scanning TEM is equipped with scanning deflection coils and detectors and mainly used to form images in the scanning mode of operation, in which the electron beam is focused to a finely condensed probe. This is achieved by directly addressing the scan coils to deflect the beam by a desired amount, producing a pore in a rapid, flexible and automated fashion. From a chemical perspective, the surface properties of nanopores govern the nature of their interaction with biomolecules. Modulation of these properties in a controlled manner can lead to the development of a highly versatile and biocompatible nanopore. Use of a mimicry motif, which resembles the structure of biological pores, to functionalize inorganic nanopore surfaces has been recently introduced by self-assembly of organic molecules.[62]

In conclusion, the use of nanopores as sensors in biological applications is still a work in progress. The creation of single nanopores has attracted much interest due to the ability to isolate and detect single molecules while they translocate through the highly confined channels. Here, we have outlined some of the options and techniques that are available when choosing single nanopores for macromolecular characterization, and how they have been, or can be, manufactured using NEMS technologies. Our studies have led to a greater insight into nanopore formation dynamics and advanced the ability to tailor nanopores for various applications in nanobiotechnology. However, there are numerous other fabrication methods, characterization techniques, applications, and mathematical models available. Moreover, other topics such as the mechanisms for solid-state nanopore formation, and the control and characterization of the geometry and surface chemistry of solid state nanopores are not fully understood. Nevertheless, nanopore technology will almost certainly yield a wealth of information on conductance mechanisms and dynamics in nanometer-sized pores, which will be exploited in the design of efficient structures for DNA fragment sizing and separation, and other applications in biosensing, nanofiltration and immunoisolation.

6.5 ACKNOWLEDGEMENTS

We would like to acknowledge our co-workers at Drexel University, the Rowland Institute at Harvard University, and that National Institute of Standards and Technology, who have helped in producing some of the results in this book chapter. Of particular note are Drs. David Bell, Joshua Edel, Amit Meller, Meni Wanunu, and Vincent Stanford. Special thanks should be given to Dr. Kazuyoshi Murata for his assistance in TEM tomography. MJK would like to thank Drs. Serge Lemay and Diego Krapf in the initial phase of the research for solid-state nanopore fabrication. JJK acknowledges the support in part by the National Institute of Standards and Technology and the Office of Law Enforcement Standards.

References

1. M. Zwolak and M. Di Ventra, *Rev. Mod. Phys.*, 2008, **80**, 141 – 166.
2. J. J. Kasianowicz, E. Brandin, D. Branton and D. W. Deamer, *Proc. Natl. Acad. Sci. U.S.A.*, 1996, **93**, 13770 – 13773.
3. A. F. Sauer-Budge, J. A. Nyamwanda, D. K. Lubensky and D. Branton, *Phys. Rev. Lett.*, 2003. **90**, 238101-1 – 238101-4.
4. A. Meller, L. Nivon and D. Branton, *Phys. Rev. Lett.*, 2001, **86**, 3435 – 3438.
5. A. Marziali and M. Akeson, *Annu. Rev. Biomed. Eng.*, 2001, **3**, 195 – 223.
6. J. Mathe, H. Visram, V. Viasnoff, Y. Rabin and A. Meller, *Biophys. J.*, 2004, **87**, 3205 – 3212.
7. M. Akeson, D. Branton, J. J. Kasianowicz, E. Brandin and D. W. Deamer, *Biophys. J.*, 1999, **77**, 3227 – 3233.
8. J. J. Kasianowicz, S. E. Herickson, H. H. Weetall and B. Robertson, *Anal. Chem.*, 2001, **73**, 2268 – 2272.
9. J. Li, D. Stein, C. McMullan, D. Branton, M. J. Aziz and J. A. Golovchenko, *Nature*, 2001, **412**, 166 – 169.
10. A. J. Storm, C. Storm, J. Chen, H. Zandbergen, J. Joanny and C. Dekker, *Nano Lett.*, 2005, **5**, 1193 – 1197.
11. M. J. Kim, M. Wanunu, D. C. Bell and A. Meller, *Adv. Mater.*, 2006, **18**, 3149 – 3153.
12. J. Li, M. Gershow, D. Stein, E. Brandin and J. A. Golovchenko, *Nature Mater.*, 2003, **2**, 611 – 615.
13. R. M. M. Smeets, U. F. Keyser, D. Krapf, M. Y. Wu, N. H. Dekker and C. Dekker, *Nano Lett.*, 2006, **6**, 89 – 95.
14. D. C. Bell, M. J. Kim, M. Wanunu and A. Meller, *Microscopy & Microanalysis*, 2006, **12**, 638 – 639.
15. L. Song, M. R. Hobaugh, C. Shustak, S. Cheley, H. Bayley and J. E. Gouaux, *Science*, 1996, **274**, 1859 – 1866.
16. S. M. Bezrukov, I. Vodyanoy, R. A. Brutyan and J. J. Kasianowicz, *Macromolecules*, 1996, **29**, 8517 – 8522.
17. S. E. Henrickson, M. Misakian, B. Robertson and J. J. Kasianowicz, *Phys. Rev. Lett.*, 2000, **85**, 3057 – 3060.

18. J. J. Kasianowicz, S. E. Henrickson, M. Misakian, H. H. Weetall, B. Robertson and V. Stanford, proceedings of the NATO Advanced Research Workshop, Budapest, 2002.
19. V. Stanford and J. J. Kasianowicz, proceedings of the IEEE Workshop on Genomic Signal Processing and Statistic, Raleigh, 2003.
20. V. M. Stanford and J. J. Kasianowicz, proceedings of the IEEE Workshop on Genomic Signal Processing and Statistics, Baltimore, 2004.
21. S. E. Henrickson, E. A. DiMarzio, Q. Wang, V. M. Stanford and J. J. Kasianowicz, *Biophys. J.*, submitted.
22. B. Hille, *Ionic channels of excitable membranes,* Sinauer Associates Inc., Massachusetts, 1992.
23. G. Menestrina, *J. Membr. Biol.*, 1986, **90**, 177 – 190.
24. R. O. Blaustein and A. Finkelstein, *J. Gen. Physiol.*, 1990, **96**, 905 – 919.
25. K. M. Halverson, R. G. Panchal, T. L. Nguyen, R. Gussio, S. F. Little, M. Misakian, S. Bavari and J. J. Kasianowicz, *J. Biol. Chem.*, 2005, **280**, 34056 – 34062.
26. S. M. Bezrukov and J. J. Kasianowicz, *Phys. Rev. Lett.*, 1993, **70**, 2352 – 2355.
27. J. J. Kasianowicz and S. M. Bezrukov, *Biophys. J.*, 1995, **69**, 94 – 105.
28. O. Braha, B. Walker, S. Cheley, J. J. Kasianowicz, M. R. Hobaugh, L. Song, J. E. Gouaux and H. Bayley, *Chem. Biol.*, 1997, **4**, 497 – 505.
29. J. J. Kasianowicz, D. L. Burden, L. Han, S. Cheley and H. Bayley, *Biophys. J.*, 1999, **76**, 837 – 845.
30. M. Misakian and J. J. Kasianowicz, *J. Membrane Biology*, 2003, **195**, 137 – 146.
31. P. Merzlyak, M. F. P. Capistrano, A. Valeva, J. J. Kasianowicz and O. V. Krasilnikov, *Biophys. J.*, 2005, **89**, 3059 – 3070.
32. P. Mueller, D. O. Rudin, T. Tien and W. C. Wescott, *Nature*, 1962, **194**, 979 – 980.
33. M. Montal and P. Mueller, *Proc. Natl. Acad. Sci. U. S. A.*, 1972, **65**, 3561 – 3566.
34. J. E. Gouaux, O. Braha, M. R. Hobaugh, L. Song, S. Cheley, C. Shustak and H. Bayley, *Proc. Natl. Acad. Sci. U.S.A.*, 1994, **91**, 12828 – 2831.
35. J. Drexler and C. Steinem, *J. Phys. Chem. B*, 2003, **107**, 11245 – 11254.
36. D. K. Shenoy, W. Barger, A. Singh, R. G. Panchal, M. Misakian, V. M. Stanford and J. J. Kasianowicz, *Nano Lett.*, 2005, **5**, 1181 – 1185.
37. L. Maissel, and R. Glang, *Handbook of Thin Film Technology*, McGraw-Hill, New York, 1970.
38. D. L. Smith, *Thin-Film Deposition: Principles and Practice*, McGraw-Hill, New York, 1994.
39. S. A. Campbell, *The Science and Engineering of Microelectronic Fabrication*, Oxford University Press, Oxford, 1996.
40. M. Madou, *Fundamentals of Microfabrication*, CRC Press, New York, 1997.
41. A. J. Storm, J. H. Chen, X. S. Ling, H. W. Zandbergen and C. Dekker, *Nature Mater.*, 2003, **2**, 537 – 540.
42. D. Krapf, M. Y. Wu, R. M. M. Smeets, H. W. Zandbergen, C. Dekker and S. G. Lemay, *Nano Lett.*, 2006, **6**, 105 – 109.
43. M. Y. Wu, D. Krapf, M. Zandbergen, H. Zandbergen and P. E. Batson, *Applied Physics Letters*, 2005, **87**, 113106-1 – 113106-3.
44. M. J. Kim, B. McNally, K. Murata and A. Meller, *Nanotechnology*, 2007, **18**, 205302-205306.
45. S. Wu, S. R. Park, and X. S. Ling, *Nano Lett.*, 2006, **11**, 2571 – 2576.
46. D. N. Mastronarde, *Microscopy & Microanalysis*, 2003, **9**, 1182 – 1183.
47. J. R. Kremer, D. N. Mastronarde and J. R. McIntosh, *J. Struct. Biol.*, 1996, **116**, 71 – 76.

48. K. –H. Klotz, M. Winterhalter and R. Benz, *Biochim. Biophys. Acta*, 1993, **1147**, 161 – 164.
49. O. V. Krasilnikov, R. Z. Sabirov, V. I. Ternovsky, P. G. Merzliak and J. N. Muratkhodjaev, *FEMS Microbiol. Immunol.*, 1992, **105**, 93 – 100.
50. S. M. Bezrukov and J. J. Kasianowicz, proceedings of the NATO Advanced Research Workshop, Budapest, 2002.
51. O. V. Krasilnikov, in *Sizing channel with polymers*, in *In Structure and Dynamics of Confined Polymers*, ed. J. J. Kasianowicz and D. W. Deamer, 2002, pp. 73 – 91.
52. S. M. Bezrukov and J. J. Kasianowicz, *Eur. Biophys. J.*, 1997, **26**, 471 – 476.
53. S. Kuga, *J. Chromatogr.*, 1981, **206**, 449 – 461.
54. J. W. F. Robertson, C. G. Rodrigues, V. M. Stanford, K. Rubinson, O. V. Krasilnikov and J. J. Kasianowicz, *Proc. Natl. Acad. Sci. U. S. A.*, 2007, **104**, 8207 – 8211.
55. J. J. Kasianowicz, J. W. F. Robertson, E. R. Chan, J. E. Reiner and V. M. Stanford, *Annual Review of Analytical Chemistry*, in press.
56. T. Ambjornsson, S. P. Apell, Z. Konkoli, E. DiMarzio and J. J. Kasianowicz, *J. Chem. Phys.*, 2002, **117**, 4063 – 4073.
57. W. Saenger, *Principles of nucleic acid structure*, Springer-Verlag, New York, 1984.
58. J. J. Kasianowicz, *Nature Mater.*, 2004, **3**, 355 – 356.
59. O. V. Krasilnikov, C. G. Rodrigues and S. M. Bezrukov, *Phys. Rev. Lett.*, 2006, **97**, 018301 – 018304.
60. J. B. Heng, C. Ho, T. Kim, R. Timp, A. Aksimentiev, Y. V. Grinkova, S. Sligar, K. Schulten and G. Timp, *Biophys. J.*, 2004, **87**, 2905 – 2911.
61. A. J. Storm, J. H. Chen, H. W. Zandbergen and C. Dekker, *Phys. Rev. E*, 2005, **71**, 051903-1 – 051903-10.
62. M. Wanunu and A. Meller, *Nano Lett.*, 2007, **7**, 1580 – 1585.

CHAPTER 7

Nanopore-Based Optofluidic Devices for Single Molecule Sensing

GUILLAUME A. T. CHANSIN[1,2], JONGIN HONG[1,2], ANDREW J. DEMELLO[1] & JOSHUA B. EDEL[1,2]

1 Department of Chemistry, Imperial College London, South Kensington, London, SW7 2AZ, United Kingdom

2 Institute of Biomedical Engineering, Imperial College London, South Kensington, London, SW7 2AZ, United Kingdom

7.1 INTRODUCTION

One of the primary motivations behind the development of miniaturised analysis devices has been to create new tools for modern day genomic and genetic analysis.[1] At present, much effort is directed towards designing faster and more efficient DNA analysis devices that could potentially identify the genes responsible for specific diseases.[2-4] One increasingly popular approach is confinement and detection of single analyte molecules within nanofluidic structures. Such devices, have at least one dimension of the channel measuring less than a few hundred nanometres.[5] One of the main advantages of nanofluidics is in the ability to confine single molecules within a well defined space in order to be efficiently detected. Importantly, probing molecules at the single molecule level is essential if one wants to measure fluctuations usually lost in ensemble averaged techniques.

Although there are many approaches in combining single molecule detection with nanofluidics, a promising approach over the past decade has been in the use of nanopores. In the 1990s, much effort was made to recreate and engineer nanopores outside of the living cell.[6] Of particular interest was the α-hemolysin pore, a protein that opens a 1.5 nm-wide channel when inserted inside a lipid bilayer membrane.[7] In 1996, Kasianowicz *et al.* demonstrated the application of α-hemolysin to measure the length of single-stranded DNA molecules.[8] The authors could detect the blockage of the ionic current during the translocation of the molecules inside the channel. This milestone opened a new field of research through the motivation that nanopores could be used to sense and analyse nucleic acids[9] (ibid. Chapter 6). In recent years, different approaches have been used to create nanopores with standard micro and nano-fabrication processing techniques as opposed to a biologically driven approach.[10,11] Such devices are made using the same materials used in the semiconductor industry and thus integrate well with other top-down technologies. In this chapter, we will focus our attention on these solid-state nanopores and their potential use in fluorescence spectroscopy of single molecules.

Nanopore-based platforms are suitable for the study of single molecules as they provide a level of confinement that allows for perfect detection efficiency. This level of confinement results in an entropic barrier being overcome by the biopolymers as the molecule is unfolded when driven inside the pore. This concept can be easily explained for macromolecules such as DNA, RNA and proteins. These molecules are long chains which can adopt a large number of conformations. They can thus be modelled using polymer physics and the size of the molecule in solution can be represented by the radius of gyration R_g. In the ideal chain approximation R_g is defined by

$$R_g^{\ 2} = \frac{1}{N}\sum_{k=1}^{N}\langle r_k^{\ 2}\rangle \tag{7.1}$$

where N is the total number of monomers and r_k is the distance of monomer k from the centre of mass coordinate.[12] In a more realistic model, the solvent and the excluded volumes (from interactions between monomers) must also be taken into account. In free solution, the polymer can adopt a very large number of conformations Z. Its entropy S is defined as

$$S = k_B \ln Z \tag{7.2}$$

where k_B is the Boltzmann constant.[13] Now let us consider the polymer inside a nanopore of radius r_0. If r_0 is much smaller than R_g the spatial confinement is such that the number of available conformation decreases (this is depicted in Figure 7.1). This decrease in entropy is called the entropic barrier and is unfavourable to the passage of the molecule inside the channel. The free energy F of the polymer is defined according to,

$$F = U - TS = U - k_B T \ln Z \tag{7.3}$$

where U is the energy of the polymer and T is the absolute temperature. At any time, the polymer tries to minimise its free energy. Hence the competing effects of external forces and entropy can induce the passage of a molecule into the channel. For example, an electric field can be used to force DNA molecules into a nanofluidic channel. This has been successfully performed and applied to the analysis of DNA stretching.[14-16]

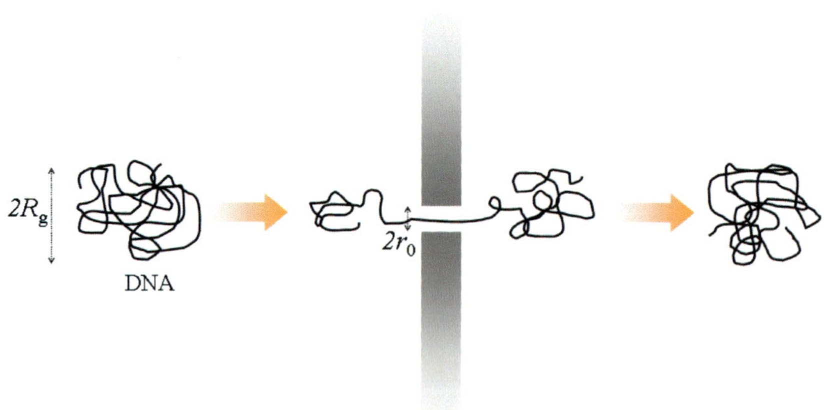

Figure 7.1 Schematic of the translocation of a DNA molecule inside a nanopore when the pore size is much smaller than the radius of gyration ($r_0 << R_g$).

It must be noted that the fabrication of a nanopore differs from that of a more conventional channel.[10] Nanopores are generally short vertical channels in a thin free-standing membrane (Figure 7.2) offering significant advantages over planar nanofluidic devices. First, the fabrication method produces nanopores with rotational symmetry. Second, the final cross sectional geometry of the nanopore does not depend on the sealing step, with pore diameters ranging from a few 100 nm to less than 5 nm,[17,18] depending on the drilling tool. Third, the relatively short channel length means the device is less prone to permanent clogging and decontamination is facilitated. Finally, the nanopores can be used as sub-wavelength apertures for optofluidic applications.

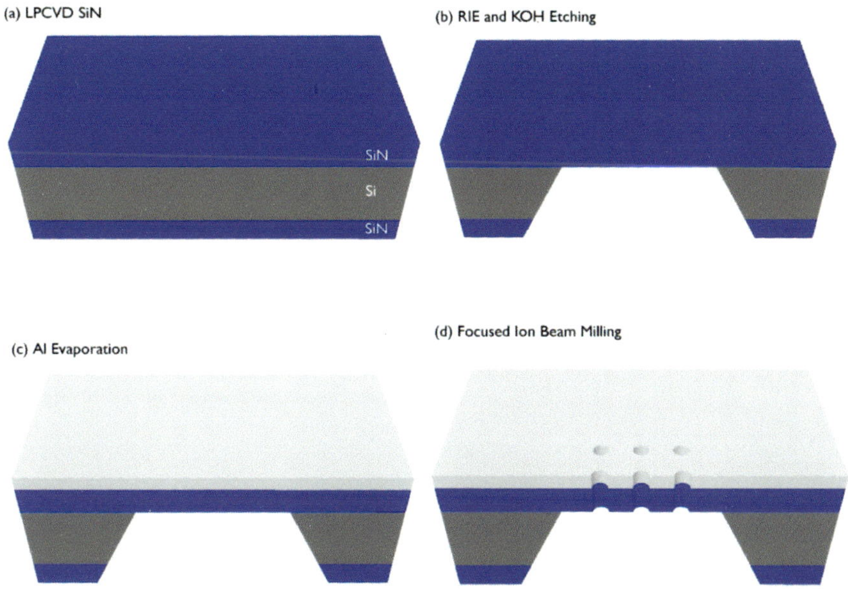

Figure 7.2 Fabrication of solid-state nanopores from a silicon wafer coated with silicon nitride (a). A window is opened in the nitride layer at the back of the wafer by standard photolithography and reactive ion etching (b). The uncovered silicon is then wet-etched with KOH to create a free standing silicon nitride membrane (b). Metal is deposited by thermal evaporation to enable the optical function of the device (c). The pores are then milled sequentially with a focused ion beam (d).

The term "optofluidic" is typically used to describe a large variety of devices that combine fluids and light.[19] These include devices containing nanoholes, or nanowells in a metal film. Unlike nanopores, these devices are not made in free standing membranes, but rather are usually fabricated on top of a glass substrate. In this chapter, we will use the term nanohole to describe any device which does not incorporate a conveying fluidic channel. Although a comprehensive review of nanohole devices is beyond the scope of this chapter, they can be categorised into two different types. The first are based on surface plasmon resonance (SPR) and usually comprise an array of nanoholes in a gold or silver film. Such devices exhibit surface plasmon-mediated extraordinary optical transmission, that results from coupling between incident radiation and surface plasmon polaritons.[20] Any change in the refractive index due to molecular interactions in the vicinity of nanohole array will induce a shift in the transmission spectrum and thus allow both real-time and label-free measurements to be performed.[21] These nanohole arrays can also be used in SPR-enhanced Raman spectroscopy.[22] A more detailed analysis of the underlying theory and application of this platform is provided by Sinton and co-workers in an excellent review article.[23]

The second class of nanohole device exploits the sub-wavelength size of a number of holes in generating extremely small probe volumes. This feature is particularly beneficial in single-molecule fluorescence spectroscopy, allowing single molecule events to be measured in samples of relatively high analytical concentration. In the current chapter, we will describe the latter category in depth and then discuss the adaptation of this format to include nanopores in a free standing membrane.

7.2 LIGHT IN SUB-WAVELENGTH PORES

Let us consider a nanohole in a metal film. How is light transmitted through such a hole if the diameter is much smaller than its wavelength? Bethe, winner of the 1967 Nobel Prize in Physics, proposed a theoretical solution to this problem. In his 1944 paper, he considered the ideal situation of a hole in a perfectly conducting metal screen with no thickness, and found a dependence of the transmitted light on the radius-to-wavelength ratio according to,

$$T \propto \left(\frac{r_0}{\lambda}\right)^4 \tag{7.4}$$

Here T is the transmission efficiency, λ the wavelength of the incident radiation and r_0 the radius of the hole in the metal screen.[24,25] This result shows that as r_0 is reduced below the wavelength, the transmission of light drops rapidly. Interestingly, there is no cut-off radius under which light propagation is forbidden. This implies that energy is transmitted even for very small hole radii. However, this theory is incomplete if the thickness of the metal screen is finite. In this case the 'hole' becomes a 'pipe' with metallic walls and is then better described as a circular waveguide.

7.2.1 Evanescent Fields in Waveguides

For a given wavelength λ, there are only a discrete number of propagation modes inside a waveguide due to symmetry as well as boundary conditions imposed by Maxwell's equations. It is common to analytically solve Maxwell's equations for two particular sets of modes: transverse-electric (TE) and transverse magnetic (TM) modes, where the direction of propagation is perpendicular to the electric field or the magnetic field, respectively. In the case of circular waveguides, these modes are written in the form TE_{pq} and TM_{pq} where p and q are integers.

Assuming r, θ, z are the cylindrical space coordinates (Figure 7.3) and t is time, each of the propagation modes has an electric field of the form (a similar equation can be written for the magnetic field)

$$\mathbf{E}_{pq}(r,\theta,z,t) = \mathbf{E}_{pq}(r,\theta)\,\mathrm{e}^{i\omega t}\,\mathrm{e}^{-\gamma z} \tag{7.5}$$

where the propagation constant γ is defined according to

$$\gamma = \pm\sqrt{\left(\frac{\chi'_{pq}}{r_0}\right)^2 - \omega^2\mu\varepsilon} = \alpha + i\beta \quad \text{for TE}_{pq} \tag{7.6}$$

$$\gamma = \pm\sqrt{\left(\frac{\chi_{pq}}{r_0}\right)^2 - \omega^2\mu\varepsilon} = \alpha + i\beta \quad \text{for TM}_{pq} \tag{7.7}$$

In these equations, μ and ε are the permeability and the permittivity of the medium filling the core of the guide, χ_{pq} is the qth root of the pth-order Bessel function, and χ'_{pq} is the qth root of the derivative of the pth-order Bessel function.[26]

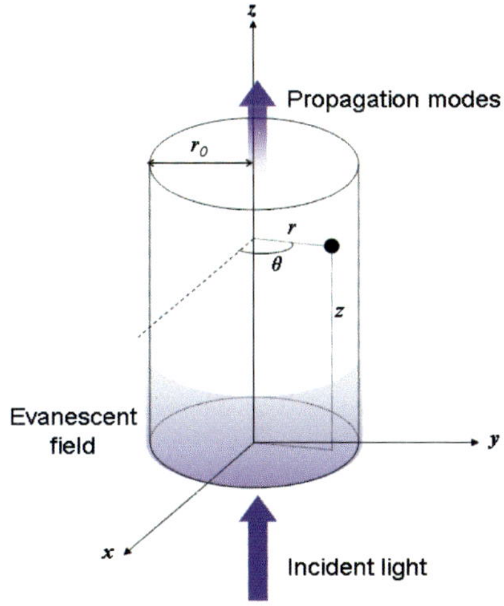

Figure 7.3 Circular waveguide in cylindrical coordinates (r, θ, z). The cladding ($r > r_0$) is made of metal and the core ($r < r_0$) is filled with a dielectric material. Modes under cut-off condition do not propagate but generate evanescent fields at the entrance of the guide.

For small radii γ is real ($\beta = 0$) and the intensity of the field decays exponentially along the z direction (an evanescent field). A cut-off condition can be defined for $\gamma = 0$ and by inserting $\lambda = 2\pi/\omega\sqrt{\mu\varepsilon}$ the cut-off diameters are given by

$$d_c = \frac{\chi'_{pq}}{\pi}\lambda \quad \text{for TE}_{pq} \tag{7.8}$$

$$d_c = \frac{\chi_{pq}}{\pi}\lambda \quad \text{for TM}_{pq} \tag{7.9}$$

For holes narrower than the cut-off diameter the mode is not allowed in the waveguide. Table 7.1 shows cut-off diameters calculated for different propagation modes. Since the wavelength depends on the refractive index of the medium inside the core of the waveguide, values are given as a percentage of λ_0 (the wavelength of the light source in vacuum). It should be noted that TE_{11} has the smallest cut-off diameter and is called the dominant mode: if the hole is narrower than its cut-off diameter, none of the other modes will be transmitted through the waveguide (i.e. the zero-mode regime). For $\lambda_0 = 488$ nm (a wavelength commonly used in optical analyses of biological systems) the zero-mode regime is obtained for diameters below 215 nm when the core is filled with water. In most optical devices this regime is usually undesirable since there is no transmission of power or data. However the evanescent field generated at the entrance of the waveguide is of interest for optofluidic applications because this phenomenon can be exploited in high resolution fluorescence detection.

Mode	TE_{01}	TM_{01}	TE_{02}	TM_{02}	TE_{11}	TM_{11}	TE_{12}	TM_{12}	TE_{21}	TM_{21}
d_c *in air*	$1.22\lambda_0$	$0.77\lambda_0$	$2.23\lambda_0$	$1.76\lambda_0$	$0.59\lambda_0$	$1.22\lambda_0$	$1.70\lambda_0$	$2.23\lambda_0$	$0.97\lambda_0$	$1.63\lambda_0$
d_c *in water*	$0.92\lambda_0$	$0.58\lambda_0$	$1.68\lambda_0$	$1.32\lambda_0$	$0.44\lambda_0$	$0.92\lambda_0$	$1.28\lambda_0$	$1.68\lambda_0$	$0.73\lambda_0$	$1.23\lambda_0$

Table 7.1 Cut-off diameters as a percentage of the operating wavelength in vacuum. Values are dependent on the refractive index of the medium in the core of the waveguide ($n = 1$ for air, $n = 1.33$ for water).

7.2.2 Zero-Mode Waveguides

Zero-mode waveguides (ZMWs) for single molecule analysis were pioneered by Craighead and coworkers.[27] These devices consist of sub-wavelength holes in a metal film in the zero-mode regime, where the short penetration depth of the evanescent field is used to minimize the optical probe volume for detection. Ultra-small probe volumes are advantageous in single-molecule detection experiments since they allow interrogation of native biological samples without the need for dilution. In other words, if the detection probe volume is kept very small, instantaneous molecular occupancies will be at or below the single molecule level even for high bulk concentrations. For example, a volume of 1 fL requires that the analyte concentration is not more than 1.6 nM. In studies by Craighead's group, ZMW excitations have been performed within aluminium films as thin as 80-100 nm.[27,28]

In the above discussion, the metal is assumed to be a perfect conductor. However, a real metal will transmit a portion of the incident electromagnetic radiation due to the inherent dispersion of the metallic dielectric function. In a typical ZMW experiment aluminium is first deposited onto a fused silica substrate and then patterned to obtain small cavities of sub-wavelength size. Fluorescence is then detected when a fluorophore reaches the bottom of the well (as is shown in Figure 7.4). Since the light is also confined by the lateral dimension of the waveguide, a detection volume of the order of attolitres to zeptolitres (10^{-18} L to 10^{-21} L) is readily achievable.[27] Compared to a conventional diffraction-limited confocal microscope this corresponds to a reduction in volume of two to five orders of magnitude. Such reduced volumes have been successfully exploited for single molecule detection at physiological concentrations[29] or for probing the surface of lipid membranes.[28,30] Furthermore, it has been reported that molecular count rates are enhanced due to a combination of shorter fluorescence lifetimes and stronger field intensities inside the nanohole.[31,32]

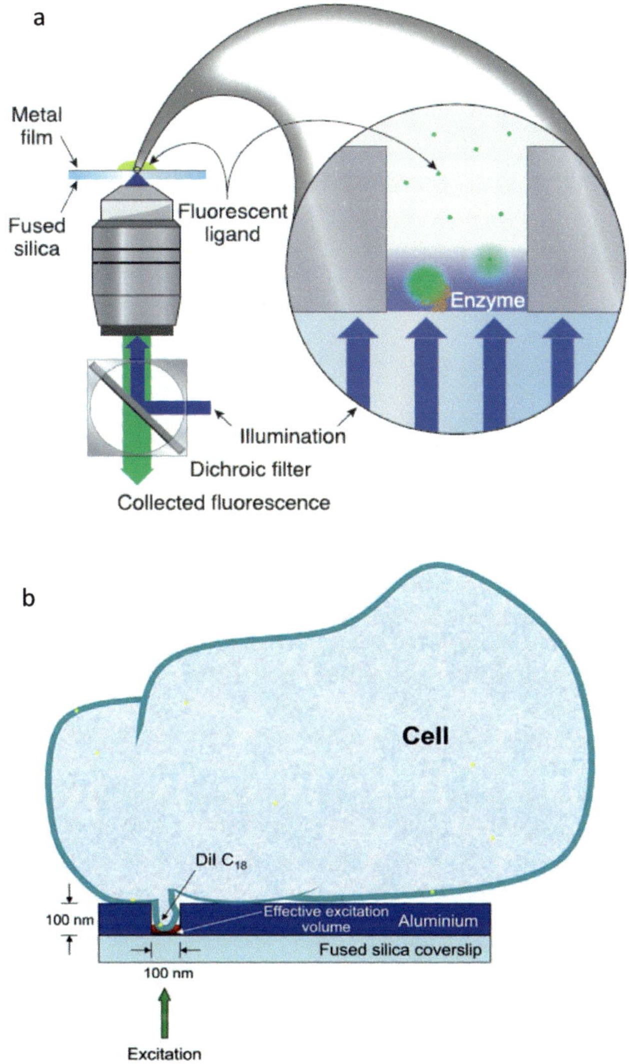

Figure 7.4 Potential applications of ZMWs. (a) Monitoring enzymatic activity. Adapted from M. J. Levene *et al.*, Science, 2003, 299, 682 and reproduced with permission. (b) Observation of single molecule along the surface of a cellular membrane. Adapted from J. B. Edel *et al.*, Biophys. J., 2005, 88, L43 and reproduced with permission.

In all experimental examples mentioned above, ZMW's have been fabricated with glass as a supporting substrate. Accordingly, analytes are not allowed to transit through the waveguide but only to diffuse in and out of the structure. However it is straightforward to notice that the geometries of a ZMW and a solid-state nanopore share certain similarities. Indeed, depending on the diameter, a cylindrical nanopore in an opaque membrane can act as a ZMW. Obtaining an optically opaque membrane is not a difficult task as several types of metal can be readily deposited on a solid-state device. The key difference when using a nanopore is the possibility of applying an external force to drive the molecules from one side of the membrane to the other. This can be done via an electrokinetic process by using a DC voltage to induce an ionic current in solution. Under these conditions, fluorescence spectroscopy becomes a viable method for probing translocations of single molecules through a pore which acts as both a fluidic channel and an optofluidic element.

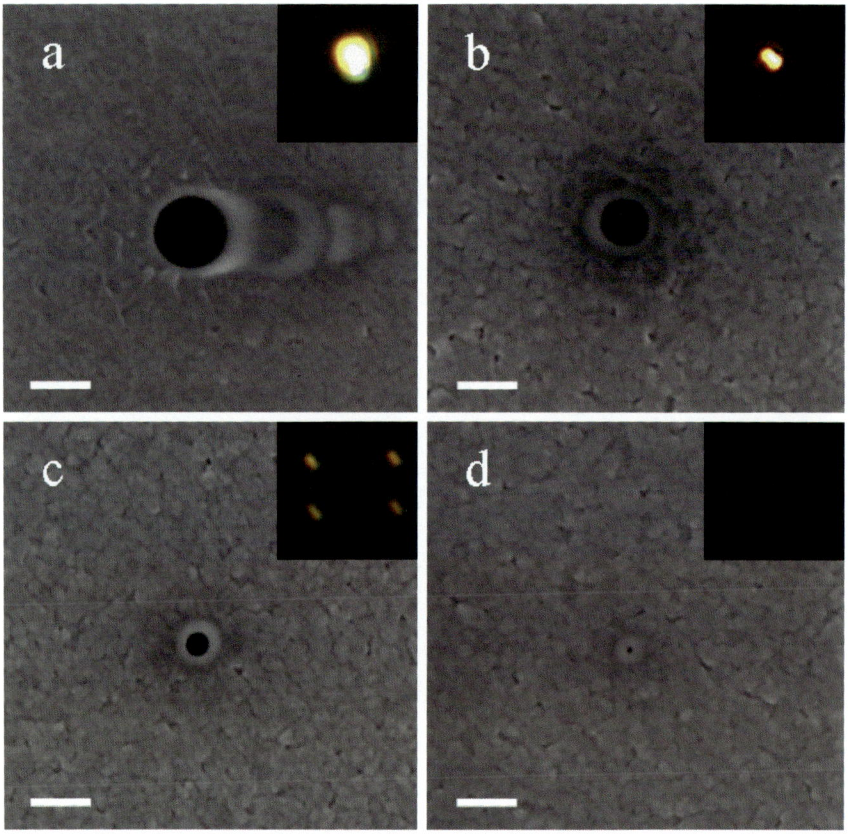

Figure 7.5 Scanning electron micrographs of nanopores in 160 nm thick membranes (60 nm SiN + 100 nm Al) with different diameters: (a) 270 nm, (b) 170 nm, (c) 100 nm, (d) 30 nm. The scale bar is 200 nm. The insets show the transmission of white light through the pores. The light is almost undetectable when the pores are 30 nm wide.

The simplest method to create a nanofluidic channel inside a solid-state membrane involves the use of focused ion beam (FIB) milling, whereby a beam of accelerated argon or gallium ions is used to remove atoms on the surface of the membrane in a physical process called sputtering. Since this method is non-selective (in terms of material preference) it can be applied to a multi-layer membrane where a metal film has been deposited on top of an initial dielectric material. Figure 7.5 shows scanning electron micrographs of FIB-milled pores of different sizes in an aluminium and silicon nitride (SiN) membrane. The integrity of the structured pores is confirmed by measuring transmitted white light using an optical microscope.

7.3 DESIGN RULES USING REAL METALS

7.3.1 Material Selection

Use of a metal layer allows suppression of light propagation through a solid-state membrane and acts as the cladding of the waveguide. As described in Figure 7.2 a thin metal film can be deposited prior to pore formation. Dielectric materials such as silicon dioxide and silicon nitride are commonly used as free-standing membranes but are optically transparent. Accordingly, it is essential that the metallic layer is thick enough to efficiently block incident light whilst not appreciably extending the nanopore length. In the waveguide theory described previously the metal cladding is considered to be a perfect conductor and the thickness of the layer has no effect on the screening efficiency. In contrast, real metals have a complex dielectric constant which can be expressed as

$$\tilde{\varepsilon}_r = \varepsilon' + i\varepsilon'' = (n + ik)^2 \tag{7.10}$$

where n is the refractive index and k is the extinction coefficient of the material. The dielectric constant depends on the incident wavelength as do n and k. The variation of n and k as a function of wavelength is shown in Figure 7.6. Figure 7.6(b) shows that metals possess large extinction coefficients and are able to efficiently attenuate visible light.

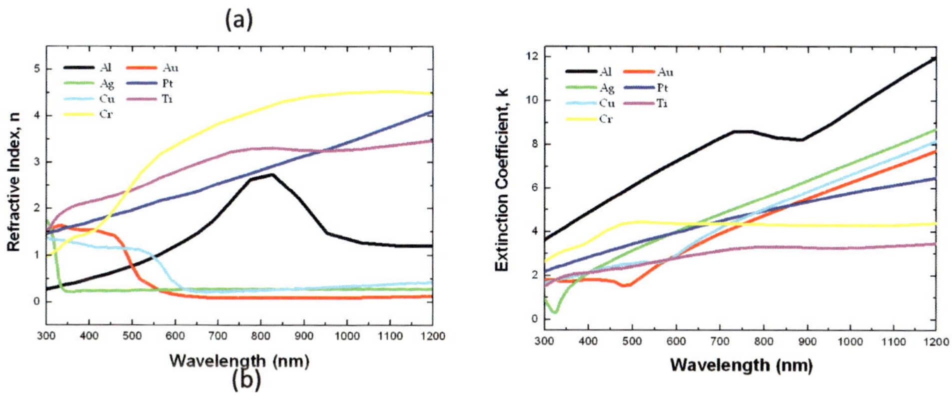

Figure 7.6 (a) Variation of refractive index as a function of wavelength and (b) variation of extinction coefficient as a function of wavelength for a selection of metals.

When detecting translocation events the signal-to-noise ratio can be maximized by ensuring that all background fluorescence from the bulk sample is removed. Accordingly, the metallic layer is crucial in efficiently screening incident light at the wavelength of interest. Two figures of merit are useful when selecting an appropriate metal for such a purpose. First, the reflectivity which is defined as the ratio of the intensity of the radiation reflected to the intensity of the incident radiation. Second, the transmission which is defined as the fraction of incident radiation that passes through the membrane. The reflectivity R at normal incidence for any absorbing media can be expressed as

$$R = \frac{(n_1 - n_2)^2 + k_2^{\,2}}{(n_1 + n_2)^2 + k_2^{\,2}} \tag{7.11}$$

where the media on each side of the interface are denoted by the numerical indices *1* and *2*.[33] The reflectivities of different materials illuminated at a wavelength of 488 nm are tabulated in Table 7.2. In order to evaluate the transmission through the membrane, the thicknesses of each material is fixed. Accordingly, transmittance values for different combinations of a 100 nm thick metal layer on top of a 200 nm thick dielectric layer were obtained via frequency-domain finite-element method (FEM) simulations and are listed in Table7.3. More details on the FEM simulations performed can be found elsewhere.[34]

Among the chosen materials, it can be observed that aluminum is the best for screening incident light, with the highest reflectivity and light attenuation properties at 488 nm. Importantly, aluminum is also free from both supporting surface plasmons (trapped electromagnetic waves at a metallic/dielectric interface formed through interaction with the free electrons in metal) and localized surface plasmons (surface plasmon modes at the edges of the hole) at 488 nm since its plasmon frequency significantly differs from the excitation wavelength. This mismatch prevents the occurrence of extraordinary transmission.

	Si	*SiN*	*SiO₂*	*Ag*	*Al*	*Au*	*Cr*	*Cu*	*Pt*	*Ti*
n	4.37	2.04	1.46	0.24	0.73	0.99	2.37	1.14	1.94	1.77
k	0.08	0	0	3.01	5.94	1.57	4.39	2.53	3.37	2.37
Reflectivity	0.28	0.044	0.0023	0.89	0.90	0.33	0.62	0.51	0.53	0.38

Table 7.2 Optical properties of different dielectric and metallic material at $\lambda_0 = 488$ nm.

	No metal	*Ag*	*Al*	*Au*	*Cr*	*Cu*	*Pt*	*Ti*
Si	-7	-35	-70	-21	-53	-31	-41	-30
SiN	-0.8	-31	-68	-16	-50	-27	-38	-26
SiO_2	-0.01	-30	-68	-15	-50	-26	-38	-25

Table7.3 Transmittance (in dB) through membranes comprising a 200 nm thick dielectric and 100 nm thick metal layer. The values are calculated at $\lambda_0 = 488$ nm.

Specifically, for aluminum, the surface plasmon resonance is located at $\omega_s \approx \omega_p / \sqrt{2}$ = 10 eV or 124 nm, where ω_s is the surface plasmon frequency and ω_p is the plasmon frequency.[35] Moreover, aluminum has a lower re-surface second harmonic generation radiation efficiency than all noble metals.[36]

7.3.2 Pore Size and Probe Volume

Figure 7.7 shows electromagnetic field distributions for Al/SiN membranes containing nanopores of varying sizes at $\lambda = 488$ nm.[34] It can be observed that as the pore size increases, more energy is transmitted to the upper (screened) region. As has been seen previously there is no propagation for pore sizes below 215 nm (Figure 7.7(b) and (c)) and the intensity of the electromagnetic field decays evanescently along the pore-axis.

However, because the thickness of the metallic layer is much smaller than the excitation wavelength and the skin-depth in real metals (which reflects the penetration of the electromagnetic field inside the walls), the electromagnetic field inside a pore is only partially attenuated (Figure 7.7(d)). Above the cut-off diameter, an appreciable amount of the electric field is transmitted into the screened region as well as within the dielectric layer, where the electric field is strongly concentrated inside the pore (Figure 7.7(e) and (f)). Both TE_{11} and TM_{01} modes are supported in the 300 nm wide pore and the TE_{21} mode is supported in the 400 nm wide pore. In addition, guided lateral propagation within the dielectric layer can be observed.

In order to accurately time translocation events through a nanopore, the radiation should be perfectly confined and thus both the size of a nanopore (especially below the cut-off diameter) and the thickness of a metallic layer are the most important engineering parameters to control. The use of smaller pore sizes and thicker metallic layer enables enhance light confinement, however, device optimization involves the consideration of other factors such as potential pore blockage and the difficulties associated with high aspect ratio nanopore fabrication.

7.4 IMPLEMENTATION AND INSTRUMENTATION

7.4.1 Detection with a Confocal Microscope

Detection of translocations through a single nanopore can be performed using a confocal microscope. The principles of this instrument were originally developed by Marvin Minsky in 1957. Since then, confocal microscopy has been extensively exploited in fluorescence spectroscopy and single molecule detection.[37] Figure 7.8 shows the basic schematic of a confocal microscope that can be used with in conjunction with a nanopore device. In such a setup, a laser beam is directed into the sample through a microscope objective. Emitted fluorescence photons are then collected by the same objective and sent to a detector. One of the key components of this set-up is the dichroic mirror which enables the spatial separation of the fluorescence signal from the excitation light. Essentially, an ideal dichroic mirror acts as a perfect mirror at the excitation wavelength and a transparent medium at the emission wavelength. For diffraction-limited optics, spatial resolution is dependent on the numerical aperture (NA) of the objective: the higher the NA the better the lateral optical resolution. Hence it is best to use oil or water-immersion objectives in which the NA is higher than unity. Finally, the detector must have a high quantum efficiency i.e. a high number of photoelectrons generated per incident photon. In this respect, avalanche photodiode detectors (APDs) are becoming the standard detector choice for use in low light conditions and for single molecule detection. For example, APDs operating in photon-counting mode and with a photon detection efficiency of >65% at 650 nm are commercially available.

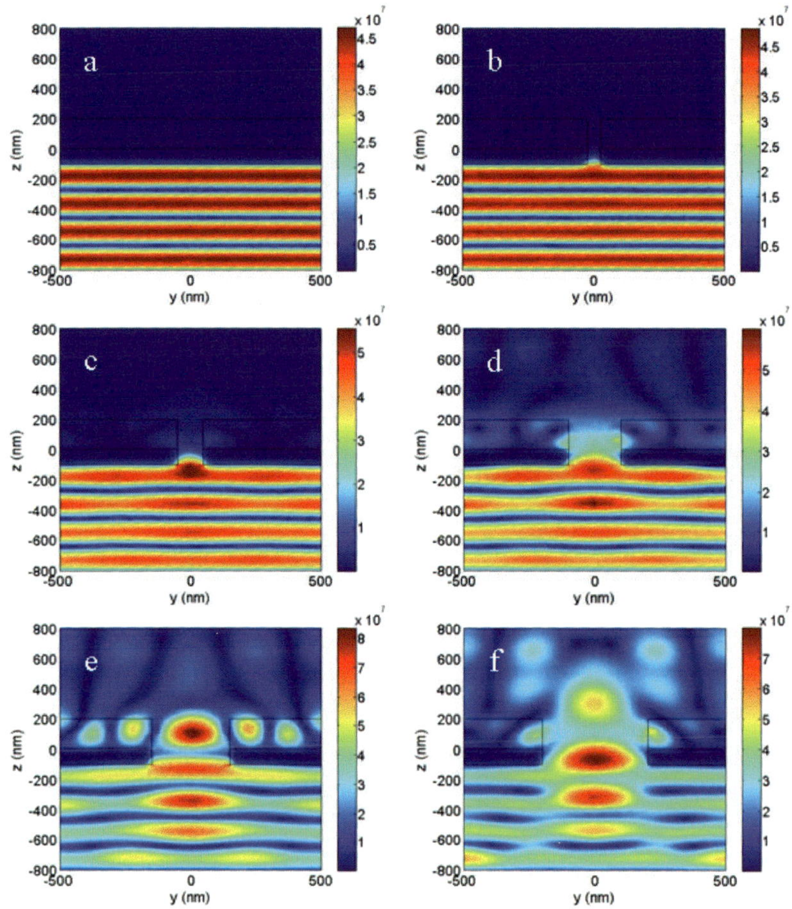

Figure 7.7 Distributions of electromagnetic field (V/m) as a function of pore diameter: (a) no hole, (b) 50 nm, (c) 100 nm, (d) 200 nm, (e) 300 nm, (f) 400 nm. Adapted from J. Hong et al., Nanotechnology, 2008, 19 and reproduced with permission.

By definition, a confocal microscope incorporates a small pinhole placed at the focal image plane.[38] The pinhole will only admit light that is focused on the aperture and rejects any out-of-focus planes. This optical sectioning provides a simple route to reducing the optical detection volume.[39] It should be noted that in the core experiments described herein a confocal pinhole is not an essential component, since the detection volume is already limited by the nanopores and the evanescent nature of the excitation light. However the pinhole is still useful at reducing the amount of fluorescent background during the measurement.

To initiate and observe translocations of molecules using the confocal microscope, the nanoporous membrane must be packaged so that it fulfils several requirements. First, the membrane must act as the main interface between the two reservoirs. Second, each reservoir must accommodate an electrode so that an electrophoretic force can be applied to the analyte molecules. Finally, the geometry of the bottom reservoir must be such that the distance between the membrane and the microscope objective is within focusing range.

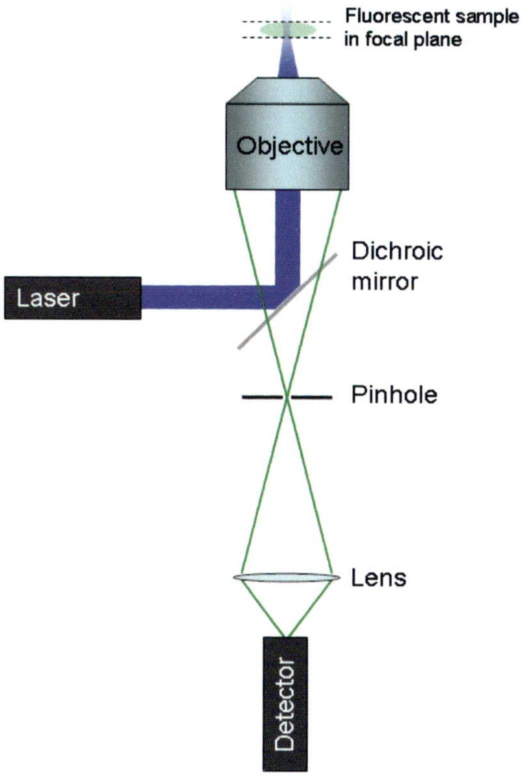

Figure 7.8 The primary features that make a confocal microscope superior to a conventional microscope are the pinhole and the objective lens. Only those rays from the best-focus point on the specimen pass through the pinhole and hit the detector. The light source is usually a laser.

Figure 7.9 depicts a schematic of the nanopore assembly where the membrane is fixed onto a glass substrate. The bottom reservoir is sealed with a glass coverslip having a thickness below 200 μm. In this situation, the coverslip is covered with a thin conductive film of indium tin oxide (ITO) on one of its surfaces. The ITO layer can then be used as the anode. A spacing layer of controllable thickness prevents any contact between the membrane and the coverslip. Figure 7.10 shows a photon burst scan of double-stranded λ-DNA (48 kbp in length) labelled with YOYO-1 as the sample. YOYO-1 is an intercalating dye and hence each DNA molecule contains more than nine thousand fluorophores along its backbone, resulting in a high signal-to-noise ratio. To generate this dataset, the laser was focused on a single pore and resulting fluorescence photons were counted using an APD. A DC voltage was applied across the membrane to drive the molecules through the pores. Accordingly, each photon burst corresponds to a DNA translocation from the top to the bottom reservoir.

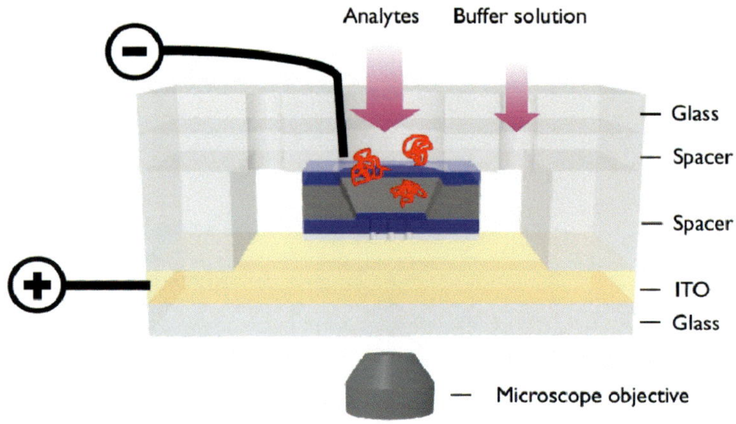

Figure 7.9 Schematic of experimental setup used to monitor DNA translocation events. Excitation light from a microscope objective is blocked by the membrane with the fluorescence signal from a translocating molecule collected by the same objective and directed to the emCCD camera. The cover slip is coated with a conductive layer of ITO. Solid-state nanopores have an average diameter of 300 nm.

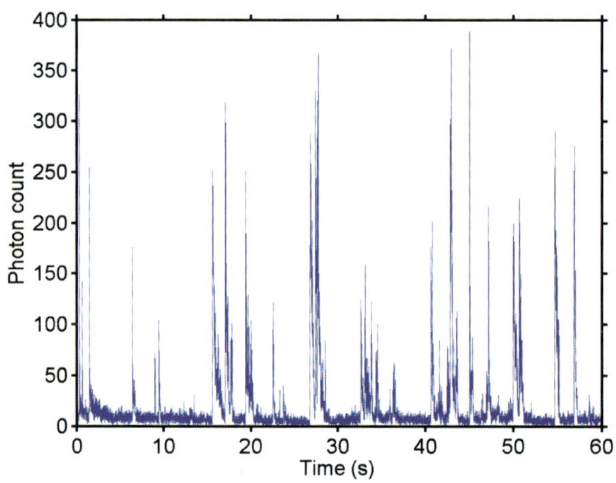

Figure 7.10 Fluorescence signal from a single pore acquired using an avalanche photodiode detector. Each peak corresponds to a single λ-DNA molecule translocating through the nanopore with a voltage of 1.60 V applied across the membrane. Data were acquired at a resolution of 50 μs and resampled at a resolution of 5 ms. Adapted from G.A.T. Chansin et al., Nano Letters, 2007, 7, 2901 and reproduced with permission.

7.4.2 Probing Nanopore Arrays using a Camera

One of the advantages of solid-state membranes is the possibility of fabricating a large array of nanopores. Electrical detection of translocations is such a format is not feasible because translocation pulses from each pore will overlap. Even if molecules could be controlled in such a way that translocations occurred simultaneously, the ionic current flowing inside the pores would not provide information on individual molecules but would result from ensemble averaging instead. Conversely, an optical setup can be readily modified to include a camera which is able to observe several pores simultaneously. It is important to note that (as with single-point detectors) the quantum efficiency of the camera must be sufficient for single-molecule detection. The frames shown in Figure 7.11 were taken with an electron-multiplying CCD (emCCD) with a quantum efficiency above 92%. The nanopores used in this experiment were 300 nm wide and are strictly speaking not ZMW.[40] However this diameter generates a stronger field intensity inside the pore, as we have seen from the simulations (Figure 7.7), and higher contrast images can be acquired.

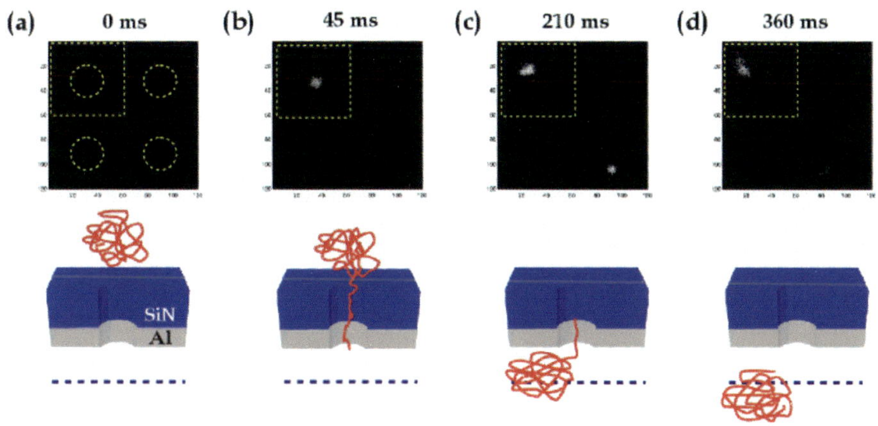

Figure 7.11 Fluorescence image of two DNA translocation events occurring under an applied voltage of 0.45 V at $t = 0$ (a), $t = 45$ ms (b), $t = 210$ ms (c), and $t = 360$ ms (d). Each pixel represents an area of 81x81 nm^2. The dotted circles in frame (a) indicate the location of the pores. Illustrations below each image frame provide an indication of the progression of DNA through the pore and the illumination plane (dotted line). Adapted from G.A.T. Chansin *et al.*, Nano Letters, 2007, 7, 2901 and reproduced with permission.

One of the primary issues associated with the use of a camera is the relatively low frame rate. For instance the measurements shown in Figure 7.11 were taken at a rate of one frame every 15 ms (67 Hz). In comparison the APD data were acquired at 20 kHz! Faster cameras are available but there will always be a compromise between speed and sensitivity. Moreover a higher resolution sensor will necessarily be slower because more pixels will need to be processed during readout. With these technological constraints it is important to ensure that the dwell time of the molecules in the probe volume is long enough for a signal to be acquired. Studies of the dwell time inside α-hemolysin pores have been reported previously. The velocity of single-stranded DNA during translocation is usually a constant but it has been shown that the velocity of a molecule shorter than the α-hemolysin channel is length-dependent.[41] Moreover, the velocity is exponentially dependent on the applied voltage[41,42] as well as on temperature.[43] Surprisingly, the velocity of double-stranded DNA in a 10 nm-wide solid state nanopore follows a very different behaviour than that of single-stranded DNA in an α-hemolysin nanopore. The duration of translocations does not grow linearly with the length of the molecule L (as expected if the velocity is constant) but scales as $L^{1.27}$.[44] From these reports it appears that, although the translocation mechanism is still not perfectly understood, voltage is the defining parameter once the molecule length, channel length and temperature have been fixed.

7.5 CONCLUSIONS

Solid-state nanopores have a great potential as tools in ultra-high throughput single-molecule analysis. Whilst most studies of nanopore systems have exploited the blockage of ionic current to observe translocations, we have seen that nanopores can also be used as optofluidic devices for high resolution fluorescence detection. Indeed by coating the solid-state membrane with an appropriate metal each nanopore can act as a waveguide where the depth and the shape of the probe volume are determined by the diameter of the pores. High-resolution confocal microscopes can be integrated with such devices to record the fluorescence signal originating from a single pore. Furthermore, unlike ionic current measurements, fluorescence can be imaged with a two-dimensional sensor, enabling the observation of a nanopore array. Probing multiple pores simultaneously has the potential to increase analytical throughput by several orders of magnitude and open paths to new applications. For example, rapid screening of a large quantity of molecules can be used for rare event detection in biological samples. Indeed, the ability to detect infrequent (< 1:10,000) cells in mixed cell populations is necessary for monitoring "minimal residual disease" in leukaemia and lymphoma patients and detecting small amounts of metastatic cells in solid tumour patients. There is no doubt that there is still much to explore, however nanopores appear to be a powerful new class of nanofluidic devices for measurements at the single molecule level.

References

1. G. M. Whitesides, *Nature*, 2006, **442**, 368-373.
2. A. J. deMello, *Nature*, 2006, **442**, 394-402.
3. D. Ryan, M. Rahimi, J. Lund, R. Mehta and B. A. Parviz, *Trends Biotechnol.*, 2007, **25**, 385-389.
4. J. O. Tegenfeldt, C. Prinz, H. Cao, R. L. Huang, R. H. Austin, S. Y. Chou, E. C. Cox and J. C. Sturm, *Anal. Bioanal. Chem.*, 2004, **378**, 1678-1692.
5. J. T. Mannion and H. G. Craighead, *Biopolymers*, 2007, **85**, 131-143.
6. H. Bayley, *Curr. Opin. Biotechnol.*, 1999, **10**, 94-103.
7. L. Z. Song, M. R. Hobaugh, C. Shustak, S. Cheley, H. Bayley and J. E. Gouaux, *Science*, 1996, **274**, 1859-1866.
8. J. J. Kasianowicz, E. Brandin, D. Branton and D. W. Deamer, *Proc. Natl. Acad. Sci. U. S. A.*, 1996, **93**, 13770-13773.
9. D. W. Deamer and M. Akeson, *Trends Biotechnol.*, 2000, **18**, 147-151.
10. M. Rhee and M. A. Burns, *Trends Biotechnol.*, 2007, **25**, 174-181.
11. C. Dekker, *Nat. Nanotechnol.*, 2007, **2**, 209-215.
12. M. Doi, *Introduction to polymer physics*, Oxford University Press Inc., New York, 1995.
13. M. Muthukumar, *Annu. Rev. Biophys. Biomolec. Struct.*, 2007, **36**, 435-450.
14. J. T. Mannion, C. H. Reccius, J. D. Cross and H. G. Craighead, *Biophys. J.*, 2006, **90**, 4538-4545.
15. C. H. Reccius, J. T. Mannion, J. D. Cross and H. G. Craighead, *Phys. Rev. Lett.*, 2005, **95**.
16. J. O. Tegenfeldt, C. Prinz, H. Cao, S. Chou, W. W. Reisner, R. Riehn, Y. M. Wang, E. C. Cox, J. C. Sturm, P. Silberzan and R. H. Austin, *Proc. Natl. Acad. Sci. U. S. A.*, 2004, **101**, 10979-10983.

17. C. Ho, R. Qiao, J. B. Heng, A. Chatterjee, R. J. Timp, N. R. Aluru and G. Timp, *Proc. Natl. Acad. Sci. U. S. A.*, 2005, **102**, 10445-10450.
18. A. J. Storm, J. H. Chen, X. S. Ling, H. W. Zandbergen and C. Dekker, *Nat. Mater.*, 2003, **2**, 537-540.
19. D. Psaltis, S. R. Quake and C. Yang, *Nature*, 2006, **442**, 381-386.
20. T. W. Ebbesen, H. J. Lezec, H. F. Ghaemi, T. Thio and P. A. Wolff, *Nature*, 1998, **391**, 667-669.
21. A. G. Brolo, R. Gordon, B. Leathem and K. L. Kavanagh, *Langmuir*, 2004, **20**, 4813-4815.
22. A. G. Brolo, E. Arctander, R. Gordon, B. Leathem and K. L. Kavanagh, *Nano Lett*, 2004, **4**, 2015-2018.
23. D. Sinton, R. Gordon and A. G. Brolo, *Microfluid. Nanofluid.*, 2008, **4**, 107-116.
24. H. A. Bethe, *Physical Review*, 1944, **66**, 163.
25. C. Genet and T. W. Ebbesen, *Nature*, 2007, **445**, 39-46.
26. J. D. Kraus, *Electromagnetics*, 4th edn., McGraw-Hill, New York, 1992.
27. M. J. Levene, J. Korlach, S. W. Turner, M. Foquet, H. G. Craighead and W. W. Webb, *Science*, 2003, **299**, 682-686.
28. J. B. Edel, M. Wu, B. Baird and H. G. Craighead, *Biophys. J.*, 2005, **88**, L43-L45.
29. K. T. Samiee, M. Foquet, L. Guo, E. C. Cox and H. G. Craighead, *Biophys. J.*, 2005, **88**, 2145-2153.
30. K. T. Samiee, J. M. Moran-Mirabal, Y. K. Cheung and H. G. Craighead, *Biophys. J.*, 2006, **90**, 3288-3299.
31. H. Rigneault, J. Capoulade, J. Dintinger, J. Wenger, N. Bonod, E. Popov, T. W. Ebbesen and P. F. Lenne, *Phys. Rev. Lett.*, 2005, **95**.
32. E. Popov, M. Neviere, J. Wenger, P. F. Lenne, H. Rigneault, P. Chaumet, N. Bonod, J. Dintinger and T. Ebbesen, *J. Opt. Soc. Am. A-Opt. Image Sci. Vis.*, 2006, **23**, 2342-2348.
33. *CRC Handbook of Chemistry and Physics*, 87th edn., Taylor and Francis, London, 2007-2008.
34. J. Hong, Y. Lee, G. A. T. Chansin, J. B. Edel and A. J. Demello, *Nanotechnology*, 2008, **19**.
35. E. Shiles, T. Sasaki, M. Inokuti and D. Y. Smith, *Phys. Rev. B*, 1980, **22**, 1612-1628.
36. D. Krause, C. W. Teplin and C. T. Rogers, *J. Appl. Phys.*, 2004, **96**, 3626-3634.
37. S. M. Nie and R. N. Zare, *Annu. Rev. Biophys. Biomolec. Struct.*, 1997, **26**, 567-596.
38. J. R. Lakowicz, *Principles of fluorescence spectroscopy*, 3rd edn., Springer, New York, 2006.
39. R. H. Webb, *Rep. Prog. Phys.*, 1996, **59**, 427-471.
40. G. A. T. Chansin, R. Mulero, J. Hong, M. J. Kim, A. J. Demello and J. B. Edel, *Nano Letters*, 2007, **7**, 2901-2906.
41. A. Meller, L. Nivon and D. Branton, *Phys. Rev. Lett.*, 2001, **86**, 3435-3438.
42. S. E. Henrickson, M. Misakian, B. Robertson and J. J. Kasianowicz, *Phys. Rev. Lett.*, 2000, **85**, 3057-3060.
43. A. Meller, L. Nivon, E. Brandin, J. Golovchenko and D. Branton, *Proc. Natl. Acad. Sci. U. S. A.*, 2000, **97**, 1079-1084.
44. A. J. Storm, C. Storm, J. H. Chen, H. Zandbergen, J. F. Joanny and C. Dekker, *Nano Lett.*, 2005, **5**, 1193-1197.

CHAPTER 8

Ion-Current Rectification in Nanofluidic Devices

LI-JING CHENG[1] & L. JAY GUO[1]

[1] Department of Electrical Engineering and Computer Science
The University of Michigan, Ann Arbor, Michigan, USA

8.1 INTRODUCTION

The transport of ions through nanoscale geometries has been widely studied due to the desire to understand the activity of biological ion channels in physiological processes and thus exploit such behaviour in biomedical and chemical applications, such as molecule delivery and sensing.[1] Fabricated on solid substrates, artificial nanochannels or nanopores provide robust and controllable environments for many applications,[2-4] and can be easily integrated with existing microfluidic devices. A key benefit of small channel sizes is the ability to handle attoliter-scale samples, resulting in minimal usage of reagent, precise quantity control and efficient processing. In addition to these benefits, nanochannels of defined size and reservoir bath concentrations offer selectivity to charged ions and molecules. This charge selectivity derives from the fact that when the size of a fluidic channel is made smaller than the Debye length, λ_D, the concentration of the counter-ions in nanochannels can be enhanced, while that of the co-ions is diminished due to the electrostatic interaction between the ions and the charged channel walls. In this regime, the electric-double layers (EDL) overlap and the ion conductance through the nanochannel is governed by the surface charge rather than the ion concentration in the reservoirs.[5,6] Such unique properties open up the possibility to selectively deliver specific types and controllable amounts of molecules or ions through the nanochannels by electrokinetic transport.

A nanofluidic device which can conduct ion current preferentially in one direction and inhibit current flow in the opposite direction is of great interest, since it may offer a new way to handle molecules or ionic species, and may even shed light on the mechanism of biological ion channels. To date, most flow controls in microfluidic devices are based on mechanical elements, such as valves and pumps. Examples include pneumatic valves actuated by gas pressure[7,8] and electrostatic valves which use electrostatic force to deform the channel structure.[9] These mechanical switches rely on the moving structures to open or close the fluidic flow and are usually relatively bulky and have slow response times. Since most biomolecules of interest are charged, it would be intriguing to explore the possibility of using regulated electrical interactions to switch their transport in a microfluidic device without any moving parts. By doing so, it may be possible to enhance the precision of flow control of ionic species and increase the integration density. In a sense, such an approach is analogous to the integrated electronic circuit, where voltage or field-controlled electronic devices, fabricated on a single substrate, control the flow of electrical current and perform a multitude of functions. To make the idea a reality, the rectifying effect is one of the essential functions to develop. Similar to the function of a diode in electronic circuits, ionic rectification allows us to turn *on* or *off* the ionic flow by simply switching the polarity of the voltage bias along the device. In this chapter, we will introduce state-of-the-art nanofluidic technologies that have been developed to produce rectifying effects in ion conduction. In addition, we will elucidate the common physics behind the rectification behaviour in different types of nanofluidic channels reported to date.

8.1.1 Analogy between Nanofluidic and Semiconductor Devices

The transport behaviour of mobile ions in electrolyte solutions shares common physics with electrons in solid state materials, especially in semiconductors. Reiss[10] and Shockley[11] pointed out that electrons and holes in semiconductors have a strong similarity to the excess and deficiency of protons in water that correspond to acidic and basic solutions. Ions and electrons resemble each other in many ways. From the aspect of transport kinetics, both ions and electrons can be basically treated using the Drude model, which describes them in the context of kinetic theory for a neutral dilute gas and hence, accounts for their Ohmic behaviour. Moreover, they both flow by diffusion and drift mechanisms. The Einstein relationship can be applied to both ions in a dilute electrolyte solution and electrons in most semiconductor materials (apart from degenerately doped materials since they obey Maxwell-Boltzmann statistics). Likewise, Debye-Hückel theory can handle both when treating the screening effect of the electric field from the individual charged particles. Additionally, the thermal generation of electrons and holes in a semiconductor is analogous to the thermal dissociation of water molecules, and weak acids and bases, in an aqueous solution. From a material point of view, it is known that electrons as majority carriers are found chiefly in an *n*-type semiconductor, where they neutralise the positively charged donors. Similarly, holes are found in *p*-type semiconductors. In nanochannels, positively and negatively charged channel surfaces play a similar role as donor and acceptor doping in *n*-type and *p*-type semiconductor materials, respectively. Nanochannels with negative surface charges have enhanced cation concentrations, while those containing positive surface charges have anions as the majority charge carriers. When two nanochannels with different surface charge polarities connect, the majority counter-ions from both sides diffuse across the junction and leave the fixed surface charge behind. The exposed surface charges generate a Donnan potential,[12] which is equivalent to the built-in potential in a semiconductor system with a non-homogenous doping distribution. The function of these potentials is exactly the same, that is, to impede the diffusion of mobile charges until equilibrium is reached with zero net charge flux. Ions at the nano-/micro-channel interface behave in the same way at equilibrium, as will be discussed in this chapter. Such equilibria, characterised by an unequal distribution of mobile ions between the two phases, are termed Donnan Equilibria.[12]

Whilst semiconductor and nanofluidic systems share some similar attributes, they are also different in several aspects. Firstly and most importantly, the majority of the cations and anions of interest do not recombine in the way electrons and holes do in semiconductors, or even as proton and hydroxyl ions do in aqueous solution. In addition, the mobilities of ions in solutions are extremely small (~ 10^{-4} cm^2/Vs) when compared to those for electrons and holes in typical semiconductors (~ 10^2-10^3 cm^2/Vs). The doping level in semiconductor materials is usually sufficient to be able to assume equal concentrations of fixed charge and majority carriers, which can be several orders of magnitude greater than that of minority carriers. However, in nanochannels, the fixed charges are located on the channel walls only, and the concentration ratio of counter-ion to co-ion is not solely determined by the surface charge density on the channel walls. Additionally, the geometry and bulk concentration of the solutions outside the nanochannels may have a strong influence as well. On the other hand, these two factors can also provide alternative ways to create asymmetry in ionic flow, and hence generate rectifying effects. This feature will be discussed subsequently.

8.2 NANOFLUIDIC DEVICES WITH RECTIFYING EFFECTS

To date, several nanofluidic devices have been shown to exhibit rectifying effects. The rectifying effects, exhibited as strong asymmetric ion conduction when switching the voltage biases, are the result of symmetry-breaking in the channel geometry, surface charge distribution, bath concentrations, or a combination of all of these aspects. We will first discuss the mechanistic details of each aspect.

8.2.1 Asymmetric Channel Geometries

Perhaps the most straightforward way to disrupt the otherwise symmetric ion conduction in a nanochannel, is to make the channel strongly asymmetric in shape. A single conical nanopore is a typical nanofluidic device that uses asymmetric geometry to rectify ion currents. Prepared by irradiating heavy ions on a 12 μm thick polymer film, such as polyethylene terephthalate (PET) or polyimide foils, and subsequent chemical etching, a single conical nanopore has a large opening diameter of ~ 600 nm on one side of the polymer membrane and a small opening diameter, ranging from 2 to 15 nm, on the other side.[13-19] The chemical etching process produces negatively charged carboxyl groups on the surfaces of these polymers, making the pore cation selective. The scanning electron microscopy (SEM) images in Figure 8.1 illustrate nanopores on PET and polyimide foils. The conical pore shape and its negative surface charge engender some interesting properties. These include the rectification of ionic current under symmetric electrolyte concentrations, and the ability to pump ions against concentration gradients driven by harmonic electric-field oscillation.[15]

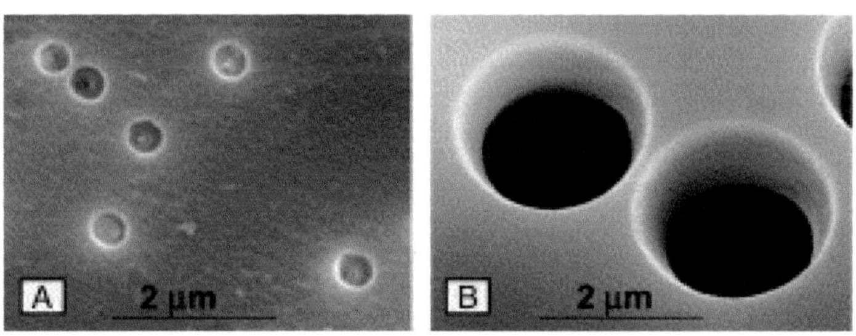

Figure 8.1 SEM images of the etched side of conical nanopores. These have a large opening on (A) a PET film irradiated by Bi ions, and (B) on a polyimide film irradiated by U ions followed by chemical etching from one side. The small pore opening is below the SEM resolution. Adapted from Z. Siwy *et al.*, Surf. Sci., 2003, 532, 1061 and reproduced with permission.

Figure 8.2 shows the typical current-voltage (I-V) characteristics of a single conical PET nanopore measured under symmetric KCl concentrations in the baths at the two ends. It can be seen that with the pore's tip side grounded, an applied negative potential on the anode gives higher ion conductance than a positive potential. It was found that the size of the pore tip must be smaller than ~ 15 nm in order to rectify the ion current at high KCl concentrations. At such a size, the radius of the pore opening at the tip side is comparable to the EDL thickness for the electrolyte concentration used. Besides geometry, it has also been found that the rectifying effect depends on the ionic strength and on the pH of the electrolyte. Rectification becomes stronger at lower KCl concentrations. The rectifying factor (or degree of rectification), which is defined as the ratio of forward bias current to reverse bias current, can be as high as 4 in a 0.1 M KCl solution at pH 7.17. Because the isoelectric point (pI) of the carboxylated pore surface is about 3, the rectifying effect exists at neutral and basic pH values, in which the carboxyl groups deprotonate leaving pore surfaces with excess negative charges. At a pH of 3, the pore surface charge becomes almost zero and the device exhibits linear I-V characteristics. This provides proof that surface charge is essential for a conical nanopore to rectify current. The surface charge of nanopores can also be modified by covering the nanopore devices with gold and then chemisorbing mercaptopropionic acid or mercaptoethylamine onto the gold surface to generate negative and positive surface charges, respectively. The I-V curves provided in Figure 8.3 show that the different polarities of the surface charge on conical nanopore devices can produce opposite rectifying I-V curves.

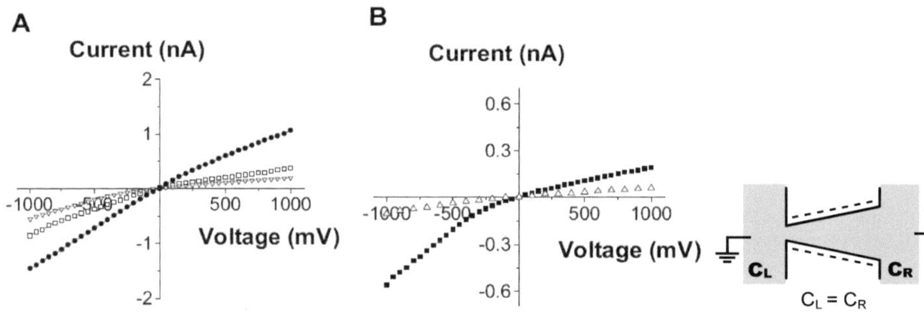

Figure 8.2 Rectifying properties of a single conical nanopore in a PET membrane with the voltage applied on the base side and the tip side grounded. (A) I-V characteristics obtained under symmetric ion concentration at pH 8 and 3 M KCl (solid circles), 1 M KCl (empty squares) and 0.1 M KCl (empty triangles). (B) I-V curves recorded under symmetric 0.1 M KCl at pH 8 (solid squares) and pH 3 (empty triangles). Adapted from Z. S. Siwy, Adv. Funct. Mater. 2006, 16, 735 and reproduced with permission.

The ionic rectifying effect has also been observed in other nanofluidic devices based on asymmetric channel geometries. Wei *et al.* pulled nanopipettes directly from clean quartz tubes, using a microprocessor-controlled CO_2 laser-based puller to shrink the opening of the glass capillaries on one side to about 20 nm.[21] As with conical nanopores, such nanopipettes possessing asymmetric channel structures and negative surface charges exhibit rectifying behaviour. In another approach, Li and Chen proposed an alternative method for the fabrication of asymmetric nanopores.[23] They exposed a 100 nm pore to an argon beam at room temperature to gradually close the pore on one side. Pore diameters as small as 1.8 nm could be achieved, and the asymmetric pore size was able to rectify ion current.

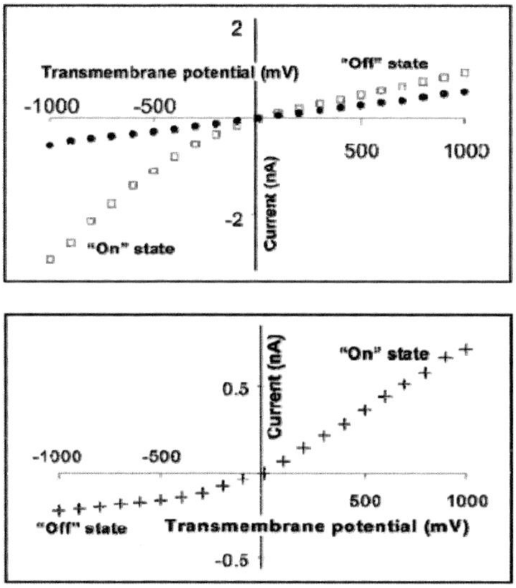

Figure 8.3 (Top) I-V curves of single gold conical nanopore modified with 2-mercaptopropionic acid to form negative surface charges, measured in 0.1 M KF, pH 6.6 (empty squares) and pH 3.5 (closed circles). (Bottom) I-V curves of a single gold conical nanopore modified with mercaptoethylammonium to form positive surface charges, obtained in 0.1 M KF, pH 6.6. Z. Adapted from Siwy *et al.*, J. Am. Chem. Soc., 2004, 126, 10850 and reproduced with permission.

8.2.2 Asymmetric Bath Concentrations

It is clear that EDL overlap is critical within a nanochannel device to control ion concentration. If such overlap varies along the channel axis in response to external potentials, the concentration of ions inside the nanochannel can be regulated accordingly. The rectifying effect can also be produced if there is a means to create an asymmetric ion distribution along a nanochannel. Based on this consideration, we have investigated ion current rectification in homogeneous nanofluidic channels placed between two reservoirs with asymmetric bath concentrations.[25] The rectifying effect is observed for a certain combination of bath KCl concentrations, such that the EDL overlaps only on one side of the channel, but not the other. The experiment was performed on planar nanochannels of two thicknesses, 20 nm and 4 nm, which were formed by removing sacrificial layers embedded in silica structures. In brief, the fabrication of nanofluidic channels started with the patterning of a silicon sacrificial layer on a glass substrate by photolithography and dry etching. After depositing a thick silicon dioxide layer to cover the sacrificial patterns, two holes were opened to expose the two ends of the sacrificial patterns by etching through the oxide layer. After removing the sacrificial layer by XeF_2 plasma etching, the channels were aligned and bonded to PDMS microfluidic channels. An O_2 plasma was then used to clean any etching residuals in the device. The channel height can be controlled by varying the thickness of the sacrificial layers. Images of this device are shown in Figure 8.4. During measurement, one bath with low concentration (termed C_L) is fixed at 0.1 mM, and the other bath with a higher concentration (termed C_H) varies from 0.1 mM to 1 M. Three regimes of I-V behaviour were observed: a symmetric regime (i.e. Ohmic behaviour), a rectifying regime, and a weakened rectifying regime.

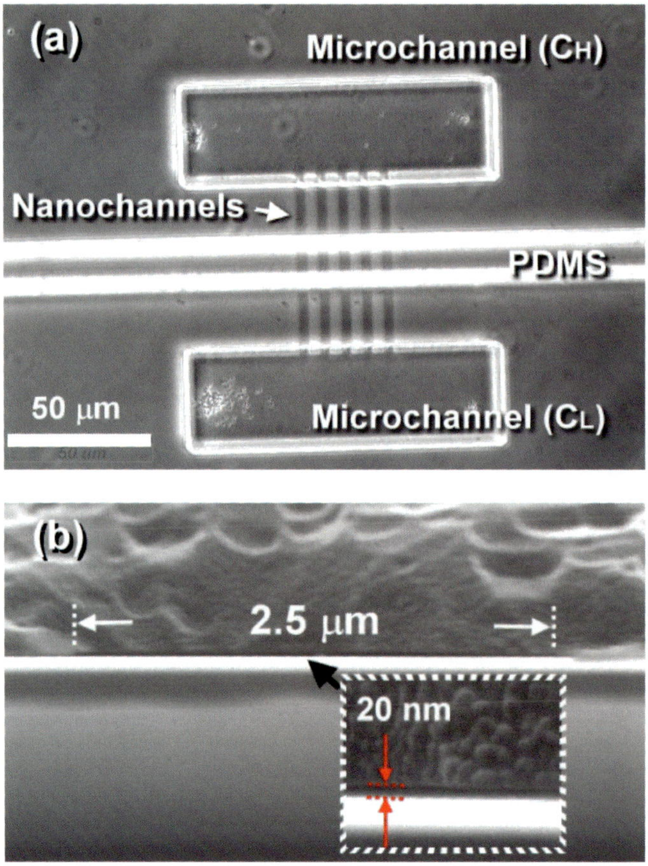

Figure 8.4 (a) Phase-contrast microscopic top-view image of the nanofluidic device. Five 60 µm long nanochannels connected with two microfluidic channels leading to the ion baths. The transparent bar across the nanochannels is a PDMS wall separating the two microchannels. (b) Cross-sectional SEM image showing that each nanochannel is 2.5 µm wide and about 20 nm thick. Adapted from L. J. Cheng and L. J. Guo, Nano Lett., 2007, 7, 3165 and reproduced with permission.

The I-V characteristics corresponding to varying C_H with a fixed C_L of 0.1 mM are plotted in Figure 8.5I(a-e) and Figure 8.5(II.a-e). Figure 8.5I(f) and Figure 8.5II(f) summarise the forward bias conductance, reverse bias conductance, and the rectifying factor I_F/I_R, where I_F and I_R are the currents measured at -5 V and 5 V bias voltage, respectively. In the symmetric regime, in which C_H is lower than 10 mM (for a 4 nm channel) or 1 mM (for a 20 nm channel), both I_F and I_R are almost identical and the nanochannels show Ohmic behaviour in conducting ionic currents. Although the reservoir concentration is asymmetric, the current level is almost equal to the results obtained in the case of symmetric concentration (i.e. $C_L = C_H = 0.1$ mM). In the second regime, the current becomes asymmetric with respect to forward and reverse voltage bias. As shown in Figure 8.5(b), the rectifying effect is maximised at $C_H = 0.1$ M with $I_F/I_R \sim 3.5$. However, in the third regime, when the C_H rises to 1 M, both I_F and I_R increase, yielding weaker asymmetric I-V characteristics. The results show that to rectify current, only one of the baths should have its bath concentration low enough to enable EDL overlap at that side of nanochannel. The bath concentration on the other side should be held high enough to avoid EDL overlap. On the other hand, the concentration on the C_H side cannot be too high; otherwise the impact of the channel surface charge on the ions and the associated rectifying effect will be weakened. These requirements for ion rectification are also applicable to conical nanopores.

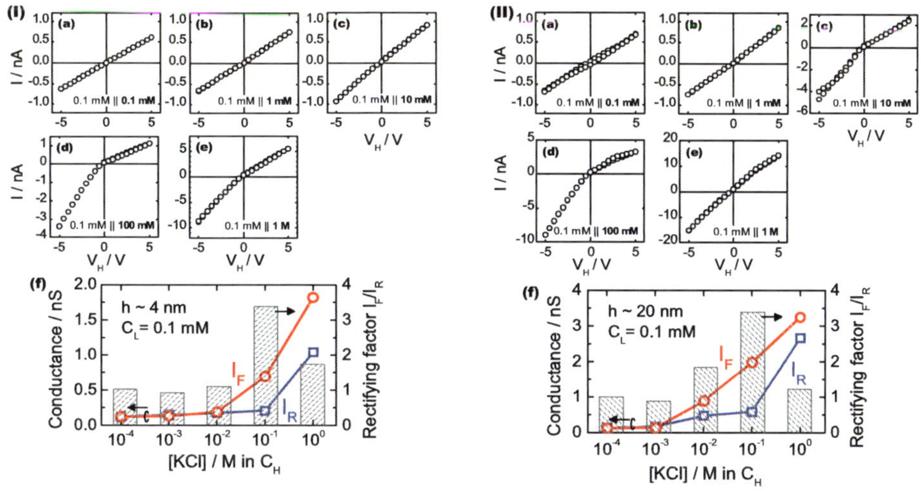

Figure 8.5 Measured I-V characteristics of 4 nm thick nanochannels (I a-e), and 20 nm thick nanochannels (II a-e) under various asymmetric concentrations ($C_L||C_H$). The forward-biased and reverse-biased conductances and the rectifying factor I_F/I_R are summarised in (I f) and (II f) for 4 nm and 20 nm thick nanochannels, respectively. The concentration in the left side (C_L) is fixed at 0.1 mM while the right side (C_H) varies from 0.1 mM to 1 M. The channel width is 2.5 μm and length, 60 μm. Adapted from L. J. Cheng and L. J. Guo, Nano Lett., 2007, 7, 3165 and reproduced with permission.

8.2.3 Asymmetric Surface Charge Distribution

A more direct way to control and produce an asymmetric ion distribution along a nanochannel is by directly creating an asymmetric surface charge. The idea of a nanofluidic diode consisting of opposite surface charges on either half of a nanochannel was proposed by Daiguji and his coworkers.[24] An asymmetric surface charge distribution in a nanochannel allows rectification of ion current. The fundamentals of such a device structure are essentially the same as the widely-studied bipolar membranes in electrochemistry. Bipolar membranes composed of a negatively charged cation exchange membrane and a positively charged anion exchange membrane present ion rectification,[25] and have been used for decades in applications such as electrodialysis and chemical separation.[26,27] Nanofluidic diodes possess similar properties but are much smaller in size and may be integrated on a fluidic chip to perform processes such as pH control and chemical separation.

The key challenge in creating a nanofluidic diode is to generate a non-uniform charge on the nanochannel surface. Karnik and co-workers have developed diffusion-limited patterning (DLP) to pattern the cationic protein avidin inside biotinylated nanofluidic channels.[28] Such nanofluidic channels were fabricated by removal of 30 nm thick silicon sacrificial channel patterns in a silica layer. The process procedures are similar to those described in Section 8.2.2. To form an asymmetric surface charge distribution, the whole surface of the 30 nm thick nanofluidic channel was first covered by biotin. Avidin was then introduced from one of the channel openings until half of the nanochannel length was covered with the avidin protein molecules.[29] Since the avidin coated and biotinylated surfaces contain positive and neutral surface charges, respectively, a surface charge discontinuity is created. Figure 8.6 shows a schematic diagram and epifluorescence image of fluorescently labelled avidin in a nanofluidic diode.

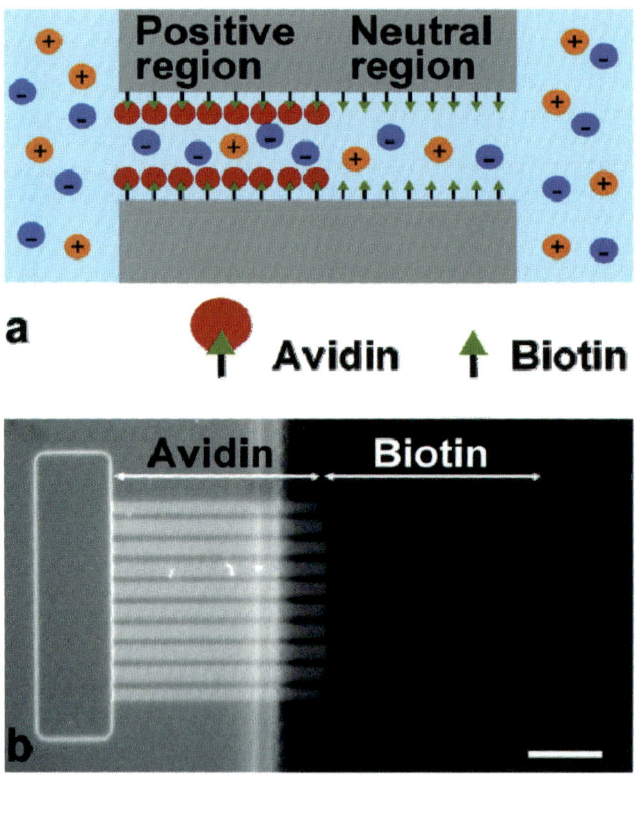

Figure 8.6 (a) Schematic diagram of a nanofluidic diode consisting of a positively charged surface and a neutral surface in different halves of the channel. The positive charge is produced by avidin, while the neutral charge is produced by biotin moieties. (b) Epifluorescence image shows fluorescently labelled avidin in the left half the nanofluidic diode. Adapted from R. Karnik *et al.*, Nano Lett., 2007, 7, 547 and reproduced with permission.

The ion rectification of the nanofluidic diode is dependent on electrolyte concentration. The experimental I-V curves under variable KCl concentrations are shown in Figure 8.7. It is seen that current rectification is prominent at intermediate concentrations of 1 mM and 10 mM KCl. The rectification effect is weaker at higher concentrations, since surface charge effects become negligible. The linear I-V characteristics were also observed at low KCl concentrations. This may be caused by enhanced water dissociation at reverse bias or polarisation at channel entrances leading to an increase of series resistance out of the nanochannel.[24,29]

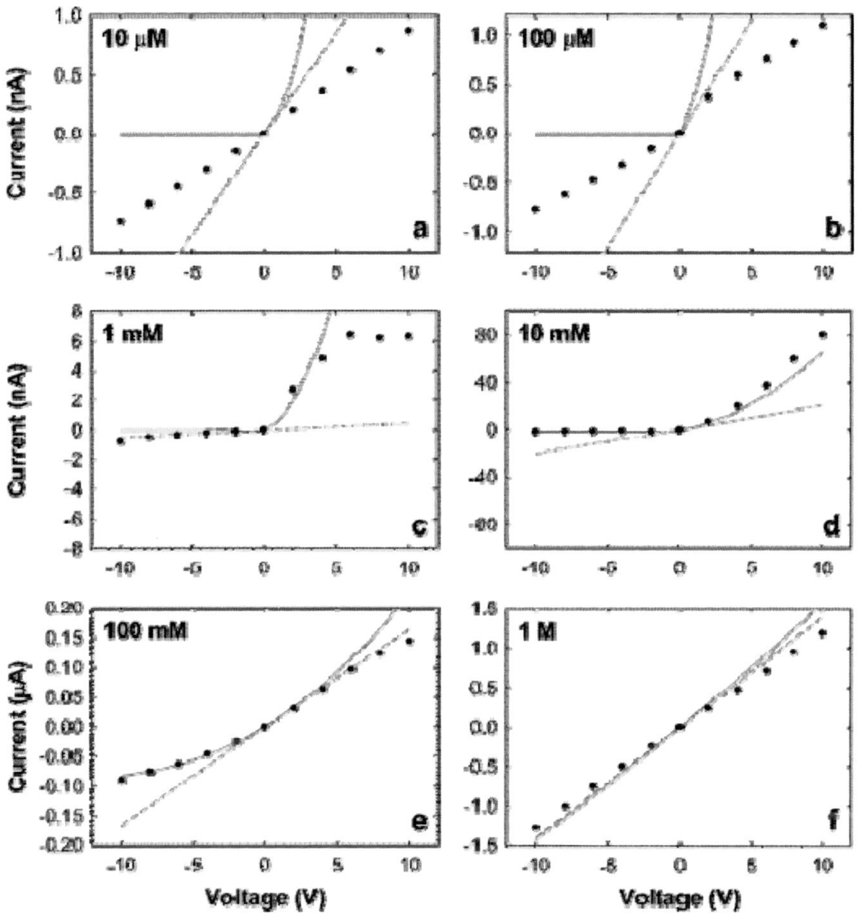

Figure 8.7 The I-V characteristics of a nanofluidic diode at different KCl concentrations (a-f). Solid circles represent experimental data, while the dashed line has slope equal to the conductance measured using a voltage bias range of -50 to 50 mV. Solid lines are theoretical predictions for an 120 μm long, 40 μm wide channel with a height of 30 nm and a surface charge of 3 mC/m^2 in the avidin half of the channel and 0 mC/m^2 in the biotin half of the channel. The biotin side is at a higher potential under forward bias. Adapted from R. Karnik *et al.*, Nano Lett., 2007, 7, 547 and reproduced with permission.

The drawback of the above surface chemical modification approach is two-fold. Firstly, the use of proteins with finite sizes may change the nanochannel dimensions and thus interfere with molecular transport. Secondly, this chemical surface modification is time consuming and insufficiently robust to allow use at temperatures other than ambient. As an alternative method, we have developed a technique to create asymmetric surface charge distributions by using different solid state oxide materials with different isoelectric points (pI). We fabricated bipolar nanochannels which incorporate silicon dioxide nanochannels and nickel oxide nanochannels connected in series.[30] The oxide nanofluidic diode relies on the precise arrangement of these oxide materials using microfabrication technology, rather than chemical treatment. Different oxide materials were patterned to sandwich part of the 20 nm thick silicon sacrificial layer. After removal of the sacrificial layer, the nanofluidic device possesses half of the channel made of silica, and the other half made of nickel oxide. The pI of a silica surface is about 3, whereas nickel oxide has a pI of between 9 and 11.29. Taking advantage of this material surface property and the high stability of the oxides, positively and negatively charged surfaces can be generated in aqueous solution over a range of pH values. Using such structures, rectification behaviour was observed with a rectifying factor of about 30 at pH 7.

Vlassiouk *et al.* reported a nanofluidic diode created by modifying the surface chemistry of conical nanopores in PET membranes.[31] The surface of the nanopore was modified to have one half of the nanopore covered by carboxyl groups and the other half covered by amino groups. The authors then studied the dependence of any rectifying behaviour on pH and electrolyte concentrations. It was found that the maximum rectifying factor occurs at a pH between 4 and 7, when the surface charge densities were maximised. Furthermore, the rectifying factor reaches 120 at 10 mM KCl, but decreases by 2 orders of magnitude when the concentration increases from 10 mM to 1 M.[31] Because the weakened rectification at higher concentration results from the change of ion transport behaviour from electrophoresis to electroosmosis, the authors conclude that it is electrophoresis that depletes ions from the nanofluidic diode at reversed bias.[31]

Finally, it is useful to note that a biological type of a nanofluidic diode has also been implemented by modifying a non-rectifying biological *porin*, *OmpF*, into a molecular diode. One way to achieve this is to mutate the protein to create the channel structure containing spatially separated selectivity filters of opposite charge.[32] A rectifying factor greater than 5 was observed in this protein p-n junction ion channel. Another way to obtain asymmetric surface charge distribution in the *porin* moiety, is to adjust the pH values in the two reservoirs. Under highly asymmetric pH conditions, the *porin* presents diode-like I-V characteristics.[33] These interesting results also imply that the rectifying effect can be achieved in a channel structure as short as the order of a Debye length.

8.3 THEORY OF RECTIFYING EFFECT IN NANOFLUIDIC DEVICES

8.3.1 Qualitative Interpretation of Ion Rectification by Solving Poisson-Nernst-Planck Equations

To understand how the rectifying effect occurs in nanofluidic channels, the ion concentration distributions and potential profiles in three types of rectifying nanofluidic devices have been calculated and reported in the literature. Several simplified models have been developed to solve the problem of ion transport in nanofluidic channels. However, all of them derive from the fundamental physical model which is based on the Poisson-Nernst-Planck (PNP) Equations,

$$-\nabla \varepsilon_0 \varepsilon_r \nabla V = F \sum_i z_i c_i \tag{8.1}$$

$$\mathbf{J}_i = -\left(D_i \nabla c_i + z_i \mu_i F c_i \nabla V\right) \tag{8.2}$$

$$\nabla \cdot \mathbf{J}_i = 0 \tag{8.3}$$

where V (V) is the electrical potential and c_i (molm^{-3}) is the concentration of ionic species i (= K^+ or Cl^-). ε_0, ε_r , F and N_A are the permittivity of vacuum, relative permittivity of the solution, the Faraday number and Avogadro's number, respectively. D_i, μ_i and z_i are the diffusivity, electrophoretic mobility and the valence number of ion species i. Equation 8.1, Poisson's equation, is used to determine the electric potential for a given charge distribution. The flux of species i is provided by Equation 8.2, the Nernst-Planck equation, which combines the diffusion current due to concentration gradients and the drift current induced by potential gradients. Since we are interested in the steady-state solution, the flux should satisfy the time-independent continuity equation, Equation 8.3. The combination of these three equations and the boundary conditions allows the calculation of ion fluxes in a system under non-equilibrium conditions.

8.3.1.1 Conical Nanopores

Cervera *et al.* computed ion concentrations and electrical potential profiles in a synthetic conical nanopore containing 0.1 M KCl responding to different voltage biases.[34] Two-dimensional PNP equations were solved in spherical coordinates with the origin at the cone apex. Since the pore opening angle is set to a small angle of 1.4°, the ionic fluxes were assumed to have only radial components. Given that the access resistance is neglected, Donnan equilibrium is assumed at both pore borders. Figure 8.8 shows the calculated results at different voltages applied across the nanopore with negative surface charges. The ion profiles indicate that at zero bias, the K^+ concentration in the nanopore is higher than the Cl^- concentration, especially close to the pore tip, due to the electrostatic interactions between charged walls and ions. When an electric field is applied from the pore tip to the base, both K^+ and Cl^- ion concentrations increase throughout the nanopore and accumulate at the tip region. As a result, the ion conductance is enlarged. On the contrary, an external electric field applied in the opposite direction depletes both types of ions and therefore, decreases the ion conductance. It should be noted that because the Donnan equilibrium is assumed at the nano-/micro-fluidic interface, information about the change of ion concentrations and the electrical potential at the pore entrance was missed.

8.3.1.2. Concentration Gradient in Homogeneous Nanochannels

We solved two-dimensional PNP equations for a system consisting of a nanochannel placed between two microchannels of different ion concentrations, which are defined as C_L and C_H for low concentration and high concentration baths, respectively.[23] Figure 8.9 illustrates the average ion concentration and potential profiles in a 10 μm long,

20 nm thick homogeneous nanochannel connecting two 2 μm square reservoirs. The left side, C_L, has a KCl concentration fixed at 0.1 mM and the right side, C_H, varies from 1 mM to 1 M KCl. Results show that at a low C_H concentration, i.e. 1 mM, at which $\mu_D \sim 10$ nm, the EDLs overlap along the entire channel. As a result, the concentration of K^+ ions is enhanced, while that of Cl^- ions is suppressed, compared to the bath concentrations (Figure 8.9(a)). When the C_H reservoir is biased, both K^+ and Cl^- ion concentrations increase slightly, but the profiles remain flat within the channel. With such low ion concentrations in both sides, the EDL overlap occurs throughout the nanochannel despite the asymmetry in the bath concentrations. Here, the ion concentrations in the nanochannel are controlled primarily by the surface, charge rather than by the applied voltages in the baths. Instead, the applied voltages induce accumulation or depletion of ions outside the nanochannel at both channel accesses. Therefore, either positive or negative bias leads to almost the same ion concentration level and potential gradient, which explains the measured symmetric I-V characteristics.

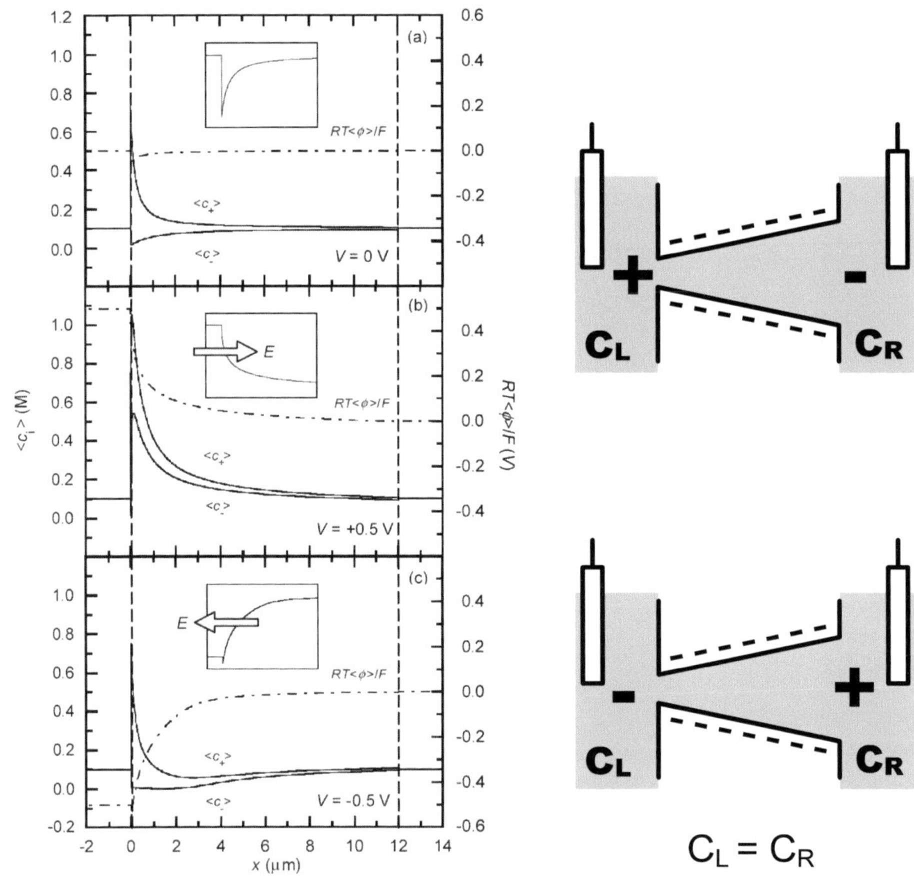

Figure 8.8 Average ion concentration distributions and electrical potential profiles in a conical nanopore calculated by PNP equations. (Top) Zero bias, (middle) 0.5 V and (bottom) -0.5 V applied on tip side, with base side grounded as illustrated in the schematic devices. In this figure, the right side bath is identified as C_R. Adapted from J. Cervera *et al.*, Europhys. Lett., 2005, 71, 35 and reproduced with permission.

As the C_H bath concentration increases to 10 mM or 100 mM, the ion concentration profile, as shown in Figure 8.9(b) and (c), either buckles up or slumps, depending on the polarities of the applied voltage. A negative bias at C_H elevates both K^+ and Cl^- concentrations in the nanochannel, leading to a high channel conductance. On the other hand, when C_H is positively-biased, both K^+ and Cl^- are depleted to a low concentration, resulting in extension of EDL overlap along the nanochannel and hence causing low channel conductance. It is this different ion distribution that produces the rectifying effect observed in this regime.

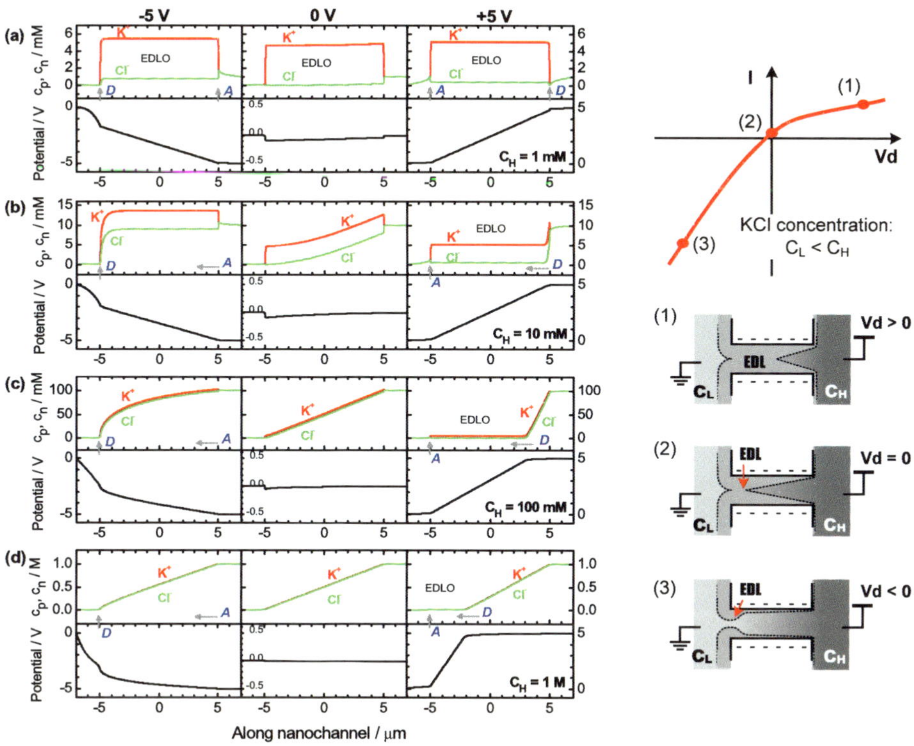

Figure 8.9 Calculated ion concentration (c_p and c_n) and potential profiles along a 20 nm thick, 10 μm long nanochannel placed between two KCl reservoirs containing different concentrations, C_L (left) and C_H (right), as illustrated in the schematics. In this system, C_L is set at ground potential and the concentration fixed to 0.1 mM, while C_H is set at (a) 1 mM, (b) 10 mM, (c) 0.1 M or (d) 1 M, and is biased at -5 V, 0 V or 5 V, as shown in the columns from left to right. The results represent the averaged values taken across the height of the nanochannel, as well as the areas which are 20 nm high in the two reservoirs extending laterally from the nanochannel. The potential profiles at zero bias are plotted with an exaggerated scale. EDLO stands for EDL overlap. The *A* or *D* with indications denotes the locations of the accumulation or depletion of ions. Adapted from L. J. Cheng and L. J. Guo, Nano Lett., 2007, 7, 3165 and reproduced with permission.

With a high concentration in C_H, 1 M for example, the number of ions is large enough to shield the surface charges in the nanochannels. In this case, almost no EDL overlap occurs throughout the entire channel, except at its very left end.

As can be seen in Figure 8.9(d), the accumulation or depletion of both K^+ and Cl^- ions is less manifest and produces a large, constant ion gradient along the channel. The reason for the weakened accumulation, when C_H is negatively-biased, is that the potential drop in the nanochannel is alleviated due to the access resistance produced by the depletion of ions around the channel access in the C_L bath. The higher the C_H, the greater the voltage drop across the access resistance in the C_L bath. In this case, the measured conductance no longer represents an intrinsic property of the nanochannel. On the other hand, at positive bias, the amount of ion depletion is comparably small, as the background ion concentration in the nanochannel increases with high C_H. Voltage biases of different polarities show less influence on ion distribution; therefore, an asymmetric I-V characteristic is not observable. Such induced access resistances should also exist in the nanopores, leading to weakened rectification at high bath concentrations.

8.3.1.3 Bipolar Nanochannels

Calculation of two-dimensional PNP equations for a nanofluidic diode which has asymmetric surface charges was first reported by Daiguji *et al.*[24] The simulation was performed for a 30 nm thick and 5 μm long nanochannel, straddled by two 1 μm square reservoirs. To understand the pressure profiles within the nanochannel, the Navier-Stokes equation was also included in the calculation.

It is assumed that the left and right halves of the channel have surface charge densities of 2 mC/m^2 and -2 mC/m^2, and that the channel contains a 5 mM KCl aqueous solution. The resulting averaged ion concentration, electrical potential and pressure profiles along the channel are plotted in Figure 8.10. They show that at forward bias, i.e. 5 V applied to the side of the negatively charged nanochannel with the other side grounded, both the majority K^+ ions in the negative channel and the majority Cl^- ions in the positive channel migrate towards the junction driven by the electric field. As a result, at steady state, both types of ions pile up around the nanochannel junction with significantly increased concentration. Thus, the high ion concentration leads to a large channel conductance.

At reverse bias, i.e. -5 V applied to the negatively charged nanochannel with the other side grounded, the counter-ions are extracted from the channel creating a depletion zone at the junction. As a result of ion depletion, channel conductance is reduced significantly. Because the flux is constant throughout the channel, the potential drops mostly in the low ion concentration regions. This explains why the potential changes gradually along a forward-biased channel, while the potential drops abruptly across the junction of a nanochannel at reverse bias.

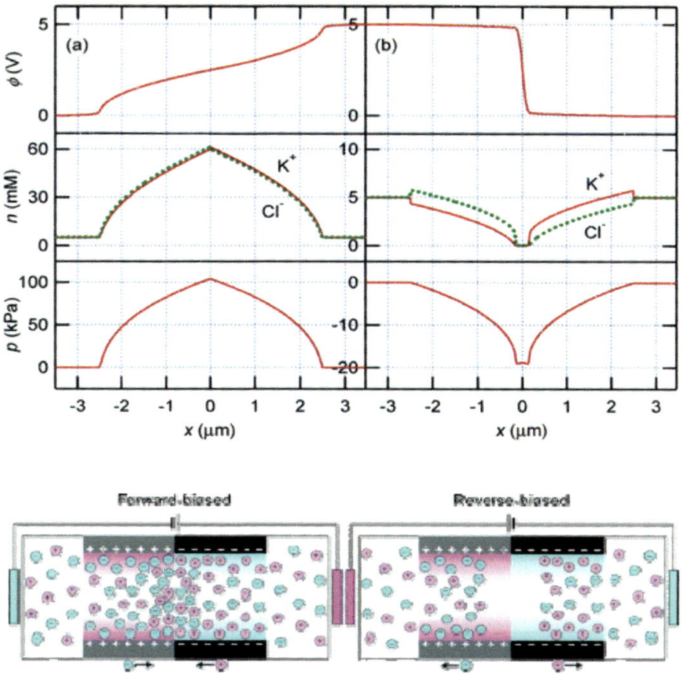

Figure 8.10 Electric potential (along channel center), ion concentration (averaged over the channel height) and pressure profiles (along channel center) for (a) a forward bias of 5 V and (b) a reverse bias of 5 V. The surface charge densities at the left and right halves of the channel are assumed to be 2 and -2 mC/m^2, respectively. The bulk concentration is 5 mM. Device schematics at the bottom show that ions accumulate at the junction under forward bias, but deplete from the junction under reverse bias. Adapted from B. Lovrecek *et al.*, Phys. Chem., 1959, 63, 750 and reproduced with permission.

8.3.2 Qualitative Interpretations of Ion Rectification in Nanofluidic Devices

Based on the calculated potential and concentration profiles obtained through the solving of PNP equations, it can be concluded that the rectifying behaviour of these three types of nanofluidic device results from the accumulation or depletion of ions in response to different bias polarities. But why do the ions inside these nanofluidic devices respond in such a similar manner? It can be envisaged that they must share some common physics. In this section, we provide a unified model to explain the rectifying effect due to the asymmetric cation/anion ratios created either by the nanochannel's asymmetric concentrations, asymmetric channel sizes or asymmetric surface charge distributions. We interpret the phenomena by analysing the drift current of K^+ and Cl^- ions near the entrances of the nanochannel immediately after application of voltage, when the space-charge that causes non-uniform electric potential has built up.[23] At this moment, ionic concentrations are still nearly identical to that at equilibrium and diffusion fluxes are negligible. Electroneutrality is assumed to be preserved in such a transient phase when ionic concentrations change over time and reach steady-state values. This method is an extension of the analysis proposed by Pu *et al.* in explaining the ion-enrichment and ion-depletion effects at low concentration at the nanochannel-microchannel interface.[35] The analysis can be broadened to explain the similar effect created inside these nanofluidic devices and hence their rectification phenomena.

Inspection of Figure 8.10(a) allows consideration of the average K^+ and Cl^- ion concentrations outside and adjacent to the right entrance of the nanochannel. At equilibrium, the concentrations of both ions are equal in the bath but unequal in the charged nanochannel, yielding the relationships of $[K]_{B,R} = [Cl]_{B,R}$ in the bath and $[K]_{n,R} = \alpha\ [Cl]_{n,R}$ in the nanochannel. Here, α is the cation/anion ratio; *B* and *n* denote the bath and the nanochannel regions, respectively, and *R*, the right hand side. In a negatively charged nanochannel, α is greater than 1, and in a positively charged nanochannel, α is less than 1. Likewise, a similar relationship, $[K]_{n,L} = \alpha\ [Cl]_{n,L}$, can be set up for the left entrance of the nanochannel. Under Donnan equilibrium, the averaged cation ($\overline{c}_+$) and anion ($\overline{c}_-$) concentrations can be calculated by solving the equation of electroneutrality $\overline{c}_+ - \overline{c}_- + f = 0$ combined with the law of mass action $\overline{c}_+ \cdot \overline{c}_- = c_b^2$, where c_b represents the bulk ion concentration outside, and *f* the nanochannel fixed charge concentration. The sign of f depends on the polarity of the fixed surface charge in the nanochannels. The calculation gives $\overline{c}_+ = -f/2 + \sqrt{(f/2)^2 + c_b^2}$ and $\overline{c}_- = f/2 + \sqrt{(f/2)^2 + c_b^2}$.[36] Thus, the cation/anion ratio, which is equal to $\overline{c}_+/\overline{c}_-$, can be expressed as

$$\alpha \text{ or } \beta = \left(\frac{-f + \sqrt{f^2 + 4\ c_b^2}}{2c_b} \right)^2, \text{ where } f = \frac{2\sigma_s 10^{-3}}{qN_A h} \tag{8.4}$$

The fixed charge concentration, *f*, is a function of surface charge density, σ_s (C/m^2), channel height, *h*, and Avogadro's number, N_A. Equation 8.4 indicates that the lower the value of c_b, or the larger the value of *f*, the greater the cation/anion ratio α (or β). A unipolar solution of counter-ions is created in the nanochannel because the surface charge of nanochannels governs the ion concentrations. When $c_b > f/2$, the cation/anion ratio decreases, as the surface charge has less influence on the ions in the nanochannel. It should be noted that the proposed equation for the cation/anion ratio is only an approximation for the case of high bath concentrations, at which nanochannels contain a non-uniform potential across their cross section.

When a negative bias is applied to the right bath, for example, both K^+ and Cl^- ions migrate in opposite directions, contributing to a drift current. By considering the current of ion *i* ($i = K^+$ or Cl^-) developed in the nanochannel ($J^i_{n,x}$) and in the bath ($J^i_{B,x}$) on the *x* side (*x* = L or R for left or right, respectively), and combining it with Kirchoff's current law (i.e. conservation of currents at a node), we obtain the relationship between the drift current in the nanochannel and the bath near the right entrance for K^+ or Cl^- ions. The current relationships are given by[35]

$$\begin{aligned} J^K_{B,R} &= \frac{\alpha^{-1} + \mu_k/\mu_{Cl}}{1 + \mu_k/\mu_{Cl}} J^K_{n,R} \\ J^{Cl}_{B,R} &= \frac{1 + \alpha\ \mu_k/\mu_{Cl}}{1 + \mu_k/\mu_{Cl}} J^{Cl}_{n,R} \end{aligned} \tag{8.5}$$

where the subscript *B,R* represents the bath and the right hand side. By imposing the approximation that the electrophoretic mobilities μ_K and μ_{Cl} are very close, Equation 8.5 is simplified to $J^K_{n,R}/J^K_{B,R} = 2/(1+\alpha^{-1})$ and $J^{Cl}_{n,R}/J^{Cl}_{B,R} = 2/(1+\alpha)$. On the basis of this derivation, we can describe the ratios of the current in the channel to that in the bath for each ion species at the left channel-bath interface, inside the nanochannel and the right channel-bath interface as following[23,37]

$$\overbrace{\begin{aligned}\frac{J^K_{n,L}}{J^K_{B,L}} &= \frac{2}{1+\beta^{-1}}\\ \frac{J^{Cl}_{n,L}}{J^{Cl}_{B,L}} &= \frac{2}{1+\beta}\end{aligned}}^{Left\ entrance}\quad \overbrace{\begin{aligned}\frac{J^K_{n,L}}{J^K_{n,R}} &= \frac{1+\alpha^{-1}}{1+\beta^{-1}}\\ \frac{J^{Cl}_{n,L}}{J^{Cl}_{n,R}} &= \frac{1+\alpha}{1+\beta}\end{aligned}}^{Nanochannel}\quad \overbrace{\begin{aligned}\frac{J^K_{n,R}}{J^K_{B,R}} &= \frac{2}{1+\alpha^{-1}}\\ \frac{J^{Cl}_{n,R}}{J^{Cl}_{B,R}} &= \frac{2}{1+\alpha}\end{aligned}}^{Right\ entrance} \tag{8.6}$$

Each current ratio listed above can be used to predict whether the ion species in that position will accumulate or deplete after an electric field is applied.

For example, let us consider the case of a negatively charged nanochannel with asymmetric bath concentrations. The left bath contains $C_L = 10^{-4}$ M. The cation/anion ratio at the left entrance, β, is calculated to be ~ 2185, when h = 20 nm and μ_s = 4.5 mC/m^2. Because the bath concentration on the right side is higher (α is small), we can get the relation: $\beta > \alpha \geq 1$. When $C_H < f/2$, both the counter-ion and co-ion currents in both ends of nanochannel are very similar ($J_{n,L} = J_{n,R}$), while those inside and outside the nanochannel are very different ($J^K_{n,R} > J^K_{B,R}$ and $J^{Cl}_{n,R} < J^{Cl}_{B,R}$). A similar result can be obtained in the C_L bath. In this case, depending on the bias polarity, the accumulation or depletion of both types of ion take place outside the nanochannel. Because the ion concentration in the nanochannel is predominately controlled by the surface charge of the channel and does not change with the bias polarity, the ion conductance is symmetric. On the contrary, when $C_H > f/2$, as depicted in Figure 8.11(1b), the ion currents in the left and right ends of the nanochannel are asymmetric ($J^K_{n,L} > J^K_{n,R}$ and $J^{Cl}_{n,L} < J^{Cl}_{n,R}$), while at the right entrance, the ion currents developed inside and outside the nanochannel are very close ($J_{n,R} = J_{B,R}$). Since the device is forward-biased (i.e. the applied electric field is opposite to the concentration gradient), as shown in Figure 8.11(1b), more ions are driven into the channel and less taken out. These uneven fluxes induce accumulation of both K^+ and Cl^- ions within the nanochannel and depletion of both near the channel entrance in the C_L bath.

Opposite results can be obtained at a reverse bias if the electric field is applied along the concentration gradient. In this situation, as illustrated in Figure 8.11(1c), more ions are being taken out of the channel than being injected into the channel, resulting in the depletion of ions in the nanochannel. The results explain why the ion concentration profiles buckle up at V_H = -5 V but sink at V_H = 5 V, as shown in Figure 8.9. In addition to the ion behaviour in the nanochannel, the accumulation and depletion of ions at the regions adjacent to the channel (i.e. the channel access region), can affect the channel conductance. The depletion of ions, especially at the channel access in the low-concentration C_L bath, increases the electrical resistance to the ion flux, and therefore generates an access resistance that is connected in series with the nanochannel. The effect of this access resistance was observed when C_H is increased to 1 M. The resistance reduces the number of ions to be accumulated or depleted in the nanochannels due to the weakened electric field inside the nanochannel and hence, reduces the rectifying effect. This analysis can be applied to a positively charged nanochannel as well. In this case, the relation of the cation/anion ratios changes to 1 $\alpha > \beta > 0$. The rectifying effect will respond to opposite bias polarities.

The nature of conical nanopores can be explained in the same way. A conical nanopore has different pore sizes, *r*, which are equivalent to different fixed charge concentrations, $f\,(= 2\,\sigma_s 10^{-3}/\,(qN_A\,r))$, on the two ends of the nanopore. Because of the small opening, the nanopore 'tip' has an equivalently higher fixed charge concentration, *f*, than its base, and therefore holds a larger cation/anion ratio than the base side (Figure 8.11(2a)). The cation/anion ratios at the two ends of the device are given by, $\beta > \alpha \geq 1$. This relation results in the asymmetric ion currents in the nanopore, $J^{K}_{n,L} > J^{K}_{n,R}$ and $J^{Cl}_{n,L} < J^{Cl}_{n,R}$. Similar to the previous conditions, at forward bias, the number of K^+ ions driven into the nanopore from the left entrance is more than those taken out through the right entrance. The Cl^- currents have similar behaviour except they move in the opposite direction. As illustrated in Figure 8.11(2b), the asymmetric flows of ions result in the accumulation of both K^+ and Cl^- in the nanopore and consequently increase the conductance. By the same token, the reverse bias depletes both types of ions out of the nanopore (Figure 8.11(2c)), yielding a low conductance.

In the case of nanofluidic diodes, a p-i nanofluidic diode,[28] which contains negative and neutral surface charges at the two sides of the nanochannel, has the condition of $\beta > \alpha = 1$; while a p-n nanofluidic diode, having negative and positive surface charges on the two sides[30,31] or a bipolar membrane, has $\beta > 1 > \alpha > 0$. As illustrated in Figure 8.11(3), when an electric field is applied to the device, the current in the negatively charged nanochannel is primarily carried by cations, whilst in the positively charged nanochannel, current is carried by anions. Since the electric field is applied from the negative nanochannel (p-type) to the positive nanochannel (n-type), the counter-ions in both sides of the nanochannel, $J^{K}_{n,L}$ and $J^{Cl}_{n,R}$, flow towards the junction.

However, the minority anions in the negative channel, $J^{Cl}_{n,L}$, and the minority cations in the positive channel, $J^{K}_{n,R}$, flow outward from the junction with smaller magnitude (Figure 8.11(3b)). As a result, both types of ions accumulate in the nanochannel and will continue to accumulate until a steady state value is reached. The accumulation of ions makes the ion concentration in the nanochannel higher than that in the baths and, therefore increases the conductance of the ions. On the contrary, if the electric field is applied in the opposite direction, ions will deplete from the junction. The depletion of ions in the nanochannel reduces its ionic conductance. As illustrated in Figure 8.11, it can be seen that having an opposite polarity of counter-ions, a biased nanofluidic diode accumulates or depletes ions in a nanochannel much more efficiently than the other two types of nanochannel. This is why nanofluidic diodes generate a higher rectifying factor than conical nanopores or nanochannels with asymmetric ion concentrations.

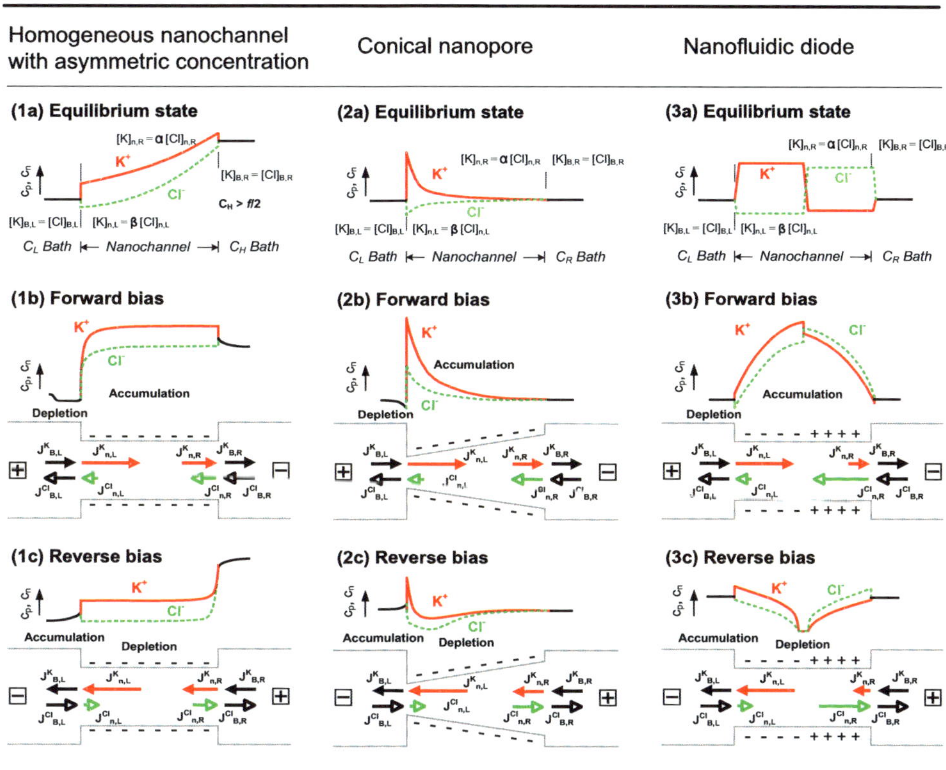

Figure 8.11 Interpretation of ionic rectification in different types of nanofluidic devices based on the analysis of symmetric ion currents building up right after the external electric fields are applied. For the same type of ion species, if the ion current flows in more than it flows out at a location, there will be an accumulation of ions when the system reaches steady state. On the contrary, if there is more out than in, depletion of ions takes place at the location.

8.3.3 Comparison of Rectifying Effects in Nanofluidic Diodes and Semiconductor Diodes

As noted previously, since ionic and electronic systems share many similarities, it would be interesting to compare the physics behind current rectification in a nanofluidic and a semiconductor diode. When an n-type and a p-type semiconductor are put together, a positive and a negative space-charge region is established at the junction, and mobile charges are depleted. The resultant built-in potential, due to the formation of space-charge, impedes the diffusion of majority carriers from both types of semiconductor across the junction. As discussed, in a nanochannel p-n diode a similar situation exists.

While both nanofluidic diodes and semiconductor diodes exhibit similar rectifying I-V characteristics, they operate differently in some ways. For instance, the forward bias current in semiconductor diodes is dominated by the diffusion of charge carriers with enough energy to overcome the built-in potential in the space-charge region. Once across the junction, these charge carriers become the minority carriers, which will recombine with the majority carriers within approximately a diffusion length from the edge of depletion or space-charge region. This result limits the potential drop in the depletion region, shielding the rest of the device from the effect of an electric field. But in a nanofluidic diode, there is no combination of a cation and an anion in the solution. Therefore, when the majority cations flow from the negative channel across the junction, driven by a forward bias, they keep flowing through the positive channel. Meanwhile, the majority anion concentrations in the positive channel must have a corresponding change in their distributions throughout the entire channel, to maintain electroneutrality. In this case, the applied electric field affects the entire device, and the drift current dominates the forward-biased ion flow.

8.4 CONCLUSIONS

Ionic rectification is a unique effect observed in nanofluidic devices but not in microfluidic devices. The rectifying phenomenon relies on the electrostatic interactions between ions and the fixed surface charges of a nanochannel. Rectification can be achieved in charged nanochannels which have asymmetric geometries, asymmetric bath concentrations or asymmetric surface charge distributions. Based on the PNP model, it is found that the rectifying phenomenon is attributed to the accumulation or depletion of ions inside nanochannels and hence, the difference in conductances in response to different bias polarities. A unified model has been developed that establishes the basis of the ionic rectifying effect in nanofluidic devices to produce asymmetric cation/anion ratios at the two entrances of the nanochannels. The condition produces the asymmetric ion currents from the two sides of the nanochannel immediately after the applied electric field builds up, and consequently allows the accumulation and depletion of ions in nanochannels to be controlled by the applied potential. Accordingly, rectifying nanofluidic devices may be used for separation and detection of charged molecules[38,39].

References

1. J. J. Kasianowicz, E. Brandin, D. Branton and D. W. Deamer, *Proc. Natl. Acad. Sci. U. S. A.*, 1996, **93**, 13770-13773.
2. T. C. Kuo, D. M. Cannon, Y. N. Chen, J. J. Tulock, M. A. Shannon, J. V. Sweedler and P. W. Bohn, *Anal. Chem.*, 2003, **75**, 1861-1867.
3. R. Karnik, K. Castelino, R. Fan, P. Yang and A. Majumdar, *Nano Lett.*, 2005, **5**, 1638-1642.
4. C. Dekker, *Nature Nanotech.*, 2007, **2**, 209-215.
5. D. Stein, M. Kruithof and C. Dekker, *Phys. Rev. Lett.*, 2004, **93**, 4.
6. H. Daiguji, P. D. Yang and A. Majumdar, *Nano Lett.*, 2004, **4**, 137-142.
7. T. M. Squires and S. R. Quake, *Rev. Mod. Phys.*, 2005, **77**, 977-1026.
8. M. A. Unger, H. P. Chou, T. Thorsen, A. Scherer and S. R. Quake, *Science*, 2000, **288**, 113-116.
9. J. Xie, J. Shih, Q. Lin, B. Yang and Y. C. Tai, *Lab Chip*, 2004, **4**, 495-501.
10. H. Reiss, *J. Chem. Phys.*, 1953, **21**, 1209-1217.
11. W. Shockley, in *Nobel Lecture*, Editon edn., 1956.
12. N. Lakshminarayanaiah, *Equations of Membrane Biophysics*, Academic Press, Orlando, 1984.

13. P. Y. Apel, Y. E. Korchev, Z. Siwy, R. Spohr and M. Yoshida, *Nucl. Instrum. Methods Phys. Res., Sect. B,* 2001, **184**, 337-346.
14. Z. Siwy, S. Mercik, K. Weron, R. Spohr, A. Wolf and Z. Grzywna, *Acta Phys. Pol., B*, 2000, **31**, 1125-1141.
15. Z. Siwy and A. Fulinski, *Phys. Rev. Lett.*, 2002, **89**, 4.
16. Z. Siwy, Y. Gu, H. A. Spohr, D. Baur, A. Wolf-Reber, R. Spohr, P. Apel and Y. E. Korchev, *Europhys. Lett.*, 2002, **60**, 349-355.
17. Z. Siwy, P. Apel, D. Dobrev, R. Neumann, R. Spohr, C. Trautmann and K. Voss, *Nucl. Instrum. Methods Phys. Res., Sect. B*, 2003, **208**, 143-148.
18. Z. Siwy, E. Heins, C. C. Harrell, P. Kohli and C. R. Martin, *J. Am. Chem. Soc.*, 2004, **126**, 10850-10851.
19. Z. Siwy, P. Apel, D. Baur, D. D. Dobrev, Y. E. Korchev, R. Neumann, R. Spohr, C. Trautmann and K. O. Voss, *Surf. Sci.*, 2003, **532**, 1061-1066.
20. Z. S. Siwy, *Adv. Funct. Mater.*, 2006, **16**, 735-746.
21. C. Wei, A. J. Bard and S. W. Feldberg, *Anal. Chem.*, 1997, **69**, 4627-4633.
22. J. Li, D. Stein, C. McMullan, D. Branton, M. J. Aziz and J. A. Golovchenko, *Nature*, 2001, **412**, 166-169.
23. L. J. Cheng and L. J. Guo, *Nano Lett.*, 2007, **7**, 3165-3171.
24. H. Daiguji, Y. Oka and K. Shirono, *Nano Lett.*, 2005, **5**, 2274-2280.
25. B. Lovrecek, A. Despic and J. O. M. Bockris, *J. Phys. Chem.*, 1959, **63**, 750-751.
26. K. N. Mani, *J. Membr. Sci.*, 1991, **58**, 117-138.
27. R. Simons and G. Khanarian, *J. Membr. Biol.*, 1978, **38**, 11-30.
28. R. Karnik, K. Castelino, C. Duan and A. Majumdar, *Nano Lett.*, 2006, **6**, 1735-1740.
29. R. Karnik, C. Duan, K. Castelino, H. Daiguji and A. Majumdar, *Nano Lett.*, 2007, **7**, 547-551.
30. L. J. Cheng and G. L. J., in *Proc. 11th MicroTAS Conf.*, Paris, France, Editon edn., 2007, pp. 979-981.
31. I. Vlassiouk and Z. S. Siwy, *Nano Lett.*, 2007, **7**, 552-556.
32. H. Miedema, M. Vrouenraets, J. Wierenga, W. Meijberg, G. Robillard and B. Eisenberg, *Nano Lett.*, 2007, **7**, 2886-2891.
33. A. Alcaraz, P. Ramirez, E. Garcia-Gimenez, M. L. Lopez, A. Andrio and V. M. Aguilella, *J. Phys. Chem. B*, 2006, **110**, 21205-21209.
34. J. Cervera, B. Schiedt and P. Ramirez, *Europhys. Lett.*, 2005, **71**, 35-41.
35. Q. S. Pu, J. S. Yun, H. Temkin and S. R. Liu, *Nano Lett.*, 2004, **4**, 1099-1103.
36. N. Lakshminarayanaiah, *Equations of Membrane Biophysics*, Academic Press, Orlando, 1984.
37. L. J. Cheng and L. J. Guo, *unpublished results*, 2008.
38. R. K. Nagarale, G. S. Gohil and V. K. Shahi, *Adv. Colloid Interface Sci.*, 2006, **119**, 97-130.
39. M. T. Gao, M. Hirata, M. Koide, H. Takanashi and T. Hano, *Process Biochem.*, 2004, **39**, 1903-1907.

CHAPTER 9

Nanopillars and Nanoballs for DNA Analysis

NORITADA KAJI[1], MANABU TOKESHI[1] & YOSHINOBU BABA[1-3]

1 Department of Applied Chemistry, MEXT Innovative Research Center for Preventive Medical Engineering, Plasma Nanotechnology Research Center, Nagoya University

2 Health Technology Research Center, National Institute of Advanced Industrial Science and Technology (AIST)

3 Institute for Molecular Science, National Institutes of Natural Sciences

9.1 INTRODUCTION

When looking back at the 1990s, it is clear that extensive research efforts were focused on the development of polymer matrices for capillary electrophoresis.[1] As has been detailed in a diversity of papers, review articles and books, the tremendous improvements in capillary electrophoresis (especially in terms of separation speed and automation) have played a key role in accelerating the Human Genome Project. In the post-genome sequencing era, the efficient use of genomic information will be key in clinical applications and the realization of personalized medicine.[2] For this reason, the analytical performance of current DNA sequencing and separation technologies must still be improved further. Microchip electrophoresis systems have been shown to be one of the strongest candidates in this respect.

Although microchip electrophoresis systems offer numerous advantages, such as high-speed, low consumption of reagents, portability and facile integration of functional components, the basic principles and derived advantages relate directly to capillary electrophoresis systems. To overcome the limitations associated with conventional electrophoresis precisely controlled nanostructures have been developed as an alternative. The pioneering work by Volkmuth and Austin in 1992 defined the concept of DNA separation using artificially fabricated structures.[3] In these studies the authors attempted to understand the motion and fractionation of large DNA molecules through beds containing well-characterized and reproducible obstacles based on the simple reptation theory developed by de Gennes.[4] With the continued development of high-efficiency micro- and nanostructuring techniques, a diversity of micro- and nanostructures have been fabricated to mimic conventional DNA sieving matrices. Significantly, these structures exhibit unusual and unique properties that are only observed in nanospace environments. Examples of such nanostructures have included consecutive shallow and deep channels for producing entropic trapping effects,[5,6] microfabricated arrays of asymmetrically arranged obstacles[7,8] and super-paramagnetic particles.[9]

In this chapter, we review and discuss our recent studies related to DNA analysis using nanofabricated structures, and more specifically the use of 'nanopillars' and 'nanoballs' for DNA size separation.[10-18]

9.2 FABRICATION OF NANOPILLARS AND NANOBALLS

Nanopillars and nanoballs have been fabricated using antithetical concepts of nanotechnology - the so-called "top-down" and "bottom-up" approaches. Both approaches offer highly effective routes to DNA analysis on the micro- and nano-scale. Recently developed nanofabrication techniques ("top-down" approaches) and polymer synthesis techniques ("bottom-up" approaches) are now discussed.

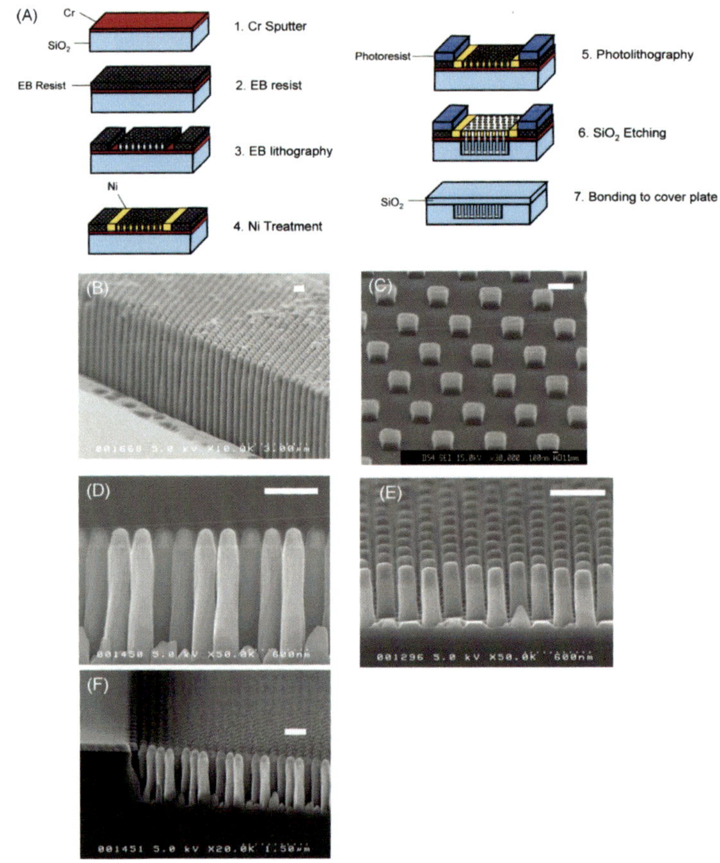

Figure 9.1 (A) Schematic description of nanofabrication processes. (B-E) Nanopillar structures fabricated on a quartz glass plate before sealing by a cover plate. The nanopillars dimensions were 200 nm wide and 4000 nm tall (aspect ratio, 20), 500 nm wide and 500 nm tall (aspect ratio, 1), 200 nm wide and 1200 nm tall (aspect ratio, 6), and 200 nm wide and 600 nm tall (aspect ratio, 3) in (B-E), respectively. (F) Nanopillar channel fabricated with Cr mask for SiO_2 dry etching. The scale bars are all 500 nm. Adapted from N. Kaji *et al.*, Anal. Chem., 76, 15, 2004 and reproduced with permission.

9.2.1 Fabrication of Nanopillars

Our method of fabricating nano-scale structures inside microfluidic channels is based on "top-down" techniques developed in conventional semiconductor fabrication industries. Although this approach requires the use of a variety of cleanroom processes, including electron-beam lithography, to delineate nanoscale patterns, precisely controlled structures with a flexible design are readily obtainable. In our nanopillar chips, the substrate material and nanopillar height are the most crucial factors in generating efficient a DNA sieving matrix based on both separation and detection requirements. Quartz is typically considered to be an excellent substrate material due to its outstanding insulating and optical transparency properties. These allow for both efficient electrophoretic separations and the facile integration of fluorescence detection schemes. To structure quartz with nanoscale features, a thin (~20 nm) Pt/Cr layer is sputtered onto a 500 μm thick quartz plate. This quartz plate is then spin-coated with a posi-type electron-beam (EB) resist (ZEP-520A). The nanopillar pattern is then delineated by EB lithography (ELS-7500, Elionix). After selective removal of the EB resist, Ni is electroplated into the holes of nanopillar pattern on the EB resist as a mask for SiO_2 dry etching and production of high aspect ratio nanopillars. Microchannels located in areas away from the nanopillar region are pattered using photolithography with a standard photoresist (OFPR-8600). Subsequently, the substrate is etched by a neutral loop discharge with CF_4. After removal of the Ni mask, the substrate is bonded to a 130 μm thick quartz cover plate with H_2SiF_6 solution at a pressure of 5 MPa. This bonding process is continued for a period of 12 hrs at 65°C. The entire process is shown schematically in Figure 9.1.

9.2.2 Self-Assembled Nanospheres

To date, natural or synthetic polymers are the first and obvious choice for DNA separations in capillary or microchip electrophoresis.[19] Although these DNA sieving matrices are readily available and have high resolving power, their viscous nature has always hindered their widespread application, especially in microchip electrophoresis. Accordingly, low-viscosity and easily-handled DNA sieving matrices are required for practical and clinical application of microchip electrophoresis.

A new type of DNA sieving matrix, based on a core-shell type of globular nanoparticle (a so-called *nanoball*) has recently been developed (Figure 9.2). These nanoballs, which consist of a block copolymer of poly(ethylene glycol) with poly(lactic acid) possessing a methacryloyl group at the PLA chain end (PEG_m-β-PLA_n-MA_l; Mw(PEG/PLA) = 6,100/4,000, m ≈ 100, n ≈ 40, l ≈ 70), tend to form small globular micelles in aqueous media. Until now, the majority of polymeric micelles have possessed no surface reactive group and the physical coagulation force of the hydrophobic core is insufficiently stable. In the described nanoball, aldehyde-PEG/PLA-methacryloyl polymer is quantitatively synthesized in order to create a stable core at the PLA end and aldehyde groups on the surface for further chemical modification. The methacryloyl groups are polymerized to form a stable core and then provided as a DNA sieving matrix.

Such nanoballs possess some remarkable properties: an average diameter of 30 nm, extremely narrow sample size distributions, no surface charge, and low viscosity in aqueous media (0.94 cP at 10 mg/ml). The viscosity of DNA sieving matrices is recognised to be a crucial factor in determining the range of potential applications.[19] Indeed, this is an especially important factor when assessing use in chip-based electrophoresis applications. Unlike conventional DNA sieving matrices which exert their sieving power through a range of pore sizes generated by inter-chain entanglements,[19] the PEG-tethered chains on the nanoball surface are thought to be the primary factor in generating the molecular sieving effect.[11]

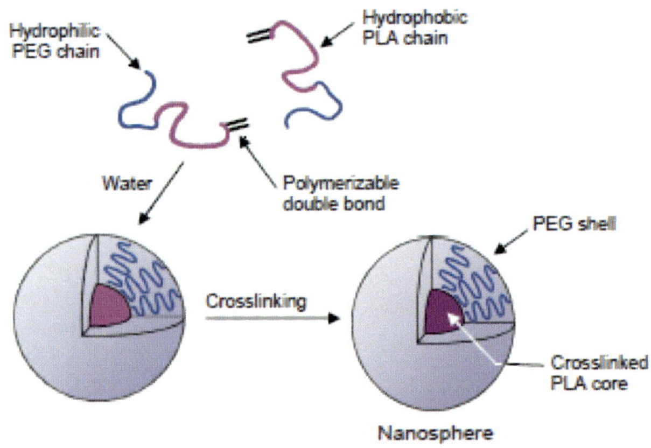

Figure 9.2 Schematic representation of the core-shell type of globular nanoparticles, so-called nanoballs. The hydrophobic PLA segments form a spherical core, which is covered by tethered, flexible PEG chains at a fairly high density. The methacryloyl groups located in the particle core are polymerized to form stable core-shell type nanospheres having a diameter of 30 nm. Adapted from M. Tabuchi, *et al.*, Nat. Biotech., 22, 337, 2004 and reproduced with permission.

9.2.3 Synthesis of PEGylated-Latex

Tabuchi *et al.* have developed another type of nanosphere based on "PEGylated-latex". This system is based on a polystyrene latex possessing poly(ethylene glycol) tethered chains on the surface.[12] Although the effectiveness of using nanoparticles as enhancement additives in DNA separations was demonstrated in 2003, gold nanoparticles (GNPs)[20] and gold nanoparticles/polymer composites (GNPPs)[21] were also shown to be electrochemically unstable during electrophoresis, exhibiting a propensity to aggregate and precipitate out of solution. To overcome such limitations, PEGylated-latex particles having a size ranging from a few nanometers to many hundreds of nanometers were developed.

A methoxy-PEG-vinylbenzene macro-monomer was synthesized via an anionic ring-opening polymerization at room temperature under an argon atmosphere using a method described elsewhere.[22] Subsequently, PEGylated latex was prepared by an emulsion polymerization of styrene in the presence of the PEG macro-monomers.[23] Using such an approach, various sizes (ranging from 80-193 nm diameter) of PEGylated-latex could be synthesized as shown in Figure 9.3. These PEGylated-latex particles were mixed into a polymer solution containing hydroxyl propyl methyl cellulose for use in DNA separations.

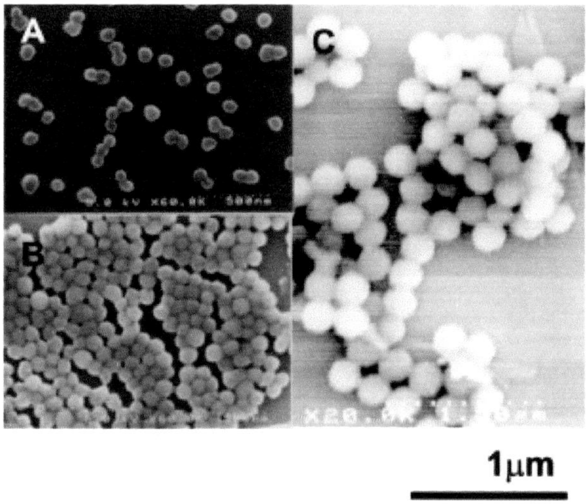

Figure 9.3 SEM image of synthesized PEGylated-latex nanoparticles having an average diameter of (A) 80 nm, (B) 110 nm, and (C) 193 nm. Adapted from M. Tabuchi *et al.*, Lab. Chip, 5, 199, 2005 and reproduced with permission

9.3 NANOPILLARS FOR DNA ANALYSIS

9.3.1 DNA Analysis by Tilted Patterned Nanopillar Chips

The first attempt to perform DNA separations using nanopillar chips was made by Kaji and co-workers in 2004.[10] In this study, the authors established a precise and controllable nanofabrication technique for constructing nanopillars on quartz substrates. The most remarkable feature of the resulting nanopillar chips was the achievable aspect ratios of the formed nanopillars. The diameter and the height of the fabricated nanopillars ranged between 100 and 500 nm and 500 and 5000 nm, respectively. Such high-aspect-ratio nanopillars are critical in increasing the available surface area not only for polymer dynamics in confined spaces but also for use as a DNA sieving matrix in microchip electrophoresis.

In this study, the authors fabricated 500 nm wide and 2700 nm high nanopillars that were arranged in a tilted array pattern along the microchannel bed. Using this type of nanopillar arrangement, several sizes of DNA fragments (between 1 and 38 kbp) and large DNA fragments (between 48.5 and 165.6 kbp) which are difficult to separate using conventional gel electrophoresis (under DC electric fields) could be successfully separated as shown in Figure 9.4. Characterisation studies resulted in separation performance of between 0.07×10^6 - 2.1×10^6 theoretical plates per meter. Although it was difficult to confirm that the separation performance using such nanopillar devices was superior to existing chip-based electrophoresis systems using conventional polymers, the results clearly showed the potential of nanopillars as DNA sieving matrices. To improve the separation performance, various factors including electroosmotic flow,[15] nanopillar array pattern,[17] and physicochemical properties of water in nanospace[13] have been carefully investigated. Some key points are now described in detail.

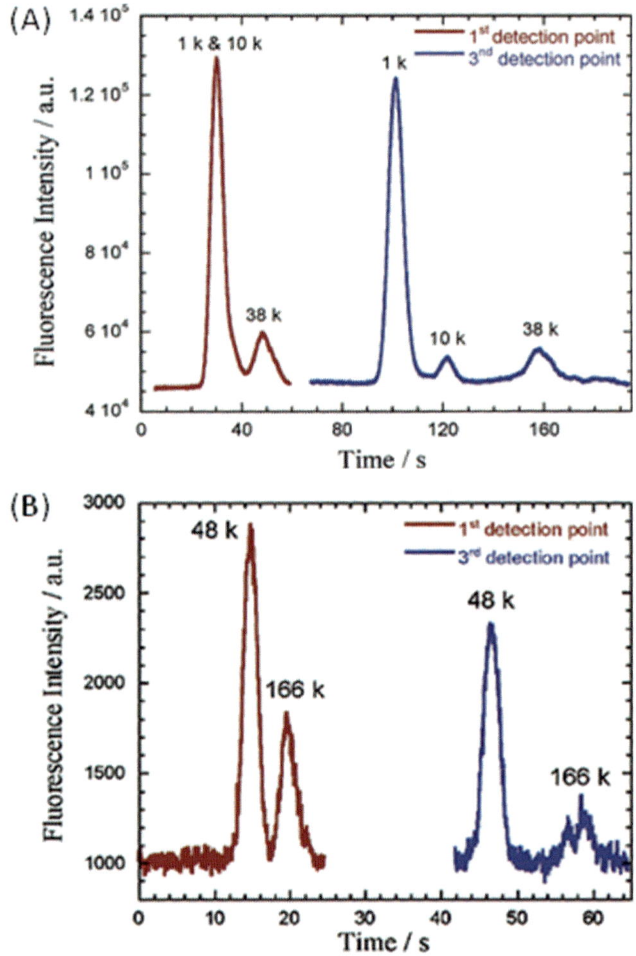

Figure 9.4 (A) Electropherograms detected at 380 (red solid line) and 1450 μm (blue solid line) from the entrance of a nanopillar channel. Separation of the mixture of 1-kbp fragments and λ–DNA digested by ApaI (10.1 and 38.4 kbp) was confirmed. The reservoir for sample loading had DNA concentrations of 1-, 10.1-, and 38.4-kbp fragments at 74, 4.2, and 16 g/mL, respectively. The peak height and width showed a good correlation with DNA concentrations. (B) Electropherograms of a λ–DNA and T4 DNA separation on the nanopillar chip at each detection point. Adapted from N. Kaji, *et al.*, Anal. Chem., 76, 15, 2004 and reproduced with permission.

9.3.2 Single DNA Molecule Imaging in Tilted Pattern Nanopillar Chips

Owing to the excellent optical properties of quartz, DNA electrophoretic migration behaviour in nanopillar chips has been clearly observed at the single molecule level.[10] Thin microchannels having a height of 2700 nm have been used to demonstrate quasi-two-dimensional DNA molecular motion. When a long DNA molecule migrates through a gel network, the DNA chain frequently gets 'hooked' on gel fibres, extending from both ends in the direction of the electric field and forming a U-shape. Here the longer arm pulls out the shorter arm, and then relaxes into a more compact conformation.[24,25] The migration of long DNA strands through tilted nanopillar chips occurs in much the same way as in a gel system.

Figure 9.5 shows the electrophoretic migration behaviour of a single λ-DNA molecule (48 kbp long) and a single T4-DNA molecule (166 kbp) within a nanopillar channel. Although the T4-DNA shows periodic conformational changes (as it does in a gel-based system), λ-DNA tends to keep a compact conformation during electrophoresis with occasional 'hook-ups' on nanopillars. These two distinct types of behaviour within the tilted patterned nanopillar chip can be understood by considering the radius of gyration of DNA. When a DNA molecule, which has larger radius of gyration than the nanopillar spacing, enters the nanopillar region, it is forced to deform to pack into the nanospace and thus will collide with the nanopillars. In contrast, smaller DNA is able to migrate through the nanopillar region with a more random-coiled conformation (mimicking its behaviour in a free solution). In the described case, the radius of gyration of λ DNA and T4 DNA is estimated to be 520 and 970 nm, respectively. Accordingly, the above interpretation could account for the observed separation of the λ DNA and T4 DNA mixture. However, it should be noted that the separation mechanism, especially from the viewpoint of polymer physics, is still poorly understood. Indeed, Yasui *et al.* have fabricated a variety of nanopillar array devices with the aim of elucidating the separation mechanism.[18]

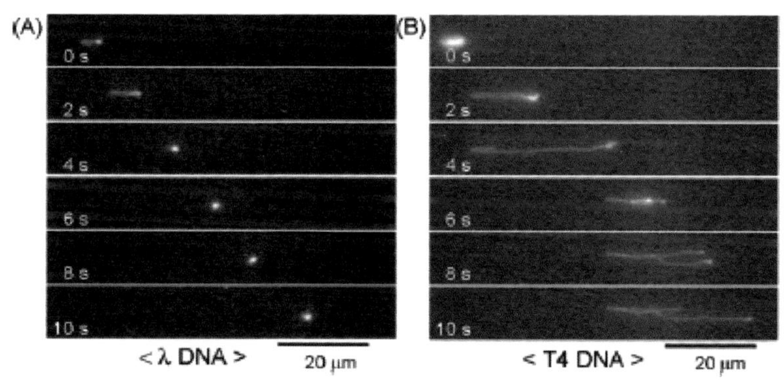

Figure 9.5 Fluorescence images of (A) a λ-DNA molecule and (B) a T4 DNA molecule migrating in a nanopillar region at 7 V/cm. Throughout a 10 second observation period, the T4 DNA appears as a U-shape formation with a greater frequency than the λ-DNA molecule. During the electrophoresis λ-DNA spends most of the time in a rather spherical conformation. Adapted from N. Kaji *et al.*, Anal. Chem., 76, 15, 2004 and reproduced with permission.)

9.3.3 DNA Analysis by Square Patterned Nanopillar Chips and Nanowall Chips

Two primary advantages associated with micro- and nanofabrication methods are design flexibility and the ability to control nanostructure feature size. To this end, another type of nanopillar array pattern, namely a square patterned nanopillar chip, has been fabricated and its separation ability studied.[17,18] Nanopillars 500 nm wide and 4000 nm high were arrayed with a 500 nm spacing with both tilted and square patterns along a microchannel (Figure 9.6). Similar sizes of DNA to those described in Section 9.3.2 were processed using the square patterned nanopillar chips. The separation results were quite unexpected and demonstrate that longer DNA molecules migrate faster than shorter DNA molecules. In the square patterned nanopillar chips, λ-DNA (48 kbp) was observed to migrate significantly more quickly than 1 kbp fragments, separating as two distinct peaks within only 15 seconds. After optimization of the separation conditions, a mixture of 1 and 5 kbp fragments could be separated within only 7 seconds. Even in the square patterned nanopillar chip, a highly-resolved and rapid separation was successfully achieved.

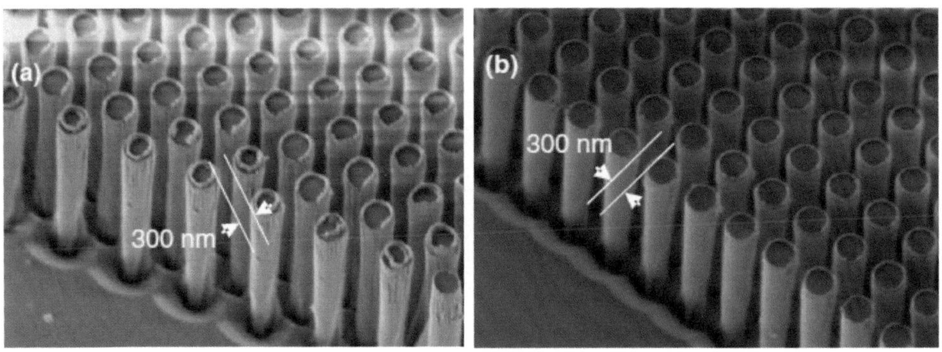

Figure 9.6 Nanopillars with a tilted distribution (a) and a square distribution (b). The gap between pillars is 300nm and the height of each nanopillar is 4 μm. Adapted from R. Ogawa, *et al.*, Jpn. J. Appl. Phys., 46, 2771, 2007 and reproduced with permission.

9.3.4 Single DNA Molecule Imaging in Square Patterned Nanopillar Chips

As in the case of the tilted patterned nanopillar chips, single DNA molecule imaging during electrophoresis in square patterned nanopillar chips has also been performed. In 500 nm-spacing nanopillar chips, T4 DNA was observed to collide with nanopillars at the interface between nanopillar and nanopillar-free region. However, after entering the nanopillar region, λ-DNA was seen to migrate in a straightforward manner without hooking the nanopillars. This migration behaviour is totally different from that observed in the tilted patterned nanopillar chips.

9.3.5 Mechanism of Separation in Nanopillar Chips

The DNA separation results observed in the square patterned nanopillar chips (i.e. longer DNA molecules migrate more quickly than shorter DNA molecules) are reminiscent of DNA separations based on entropic trapping proposed by Han and co-workers.[5,6] In this approach, the consecutive repetition of shallow and deep channel features generates free energy gaps against DNA molecules. Since the free energy gap is proportional to the size of DNA, larger DNA molecules can escape from the shallow region more quickly than shorter strands, and thus migrate at a higher average velocity. Using this analysis, the nanopillar structures described in this chapter could be regarded as analogous entropic barriers. To investigate the effect of this free energy gap at the interface of nanopillar and nanopillar-free regions, nanowall structures, aimed at excluding DNA and nanopillar interactions inside the nanopillar region, have been fabricated and provided as novel nanostructures.[18] These nanostructures are expected to reveal more detailed aspects of the DNA separation mechanism.

9.4 NANOBALLS FOR DNA ANALYSIS

9.4.1 DNA Analysis by a Self-Assembled Nanosphere Solution in a Chip

In a 1% (10 mg/ml) self-assembled nanosphere solution, a 100 bp and a 1 kbp dsDNA ladder were successfully separated within 60 and 100 seconds, respectively.[11] The results of these experiments are shown in Figure 9.7.

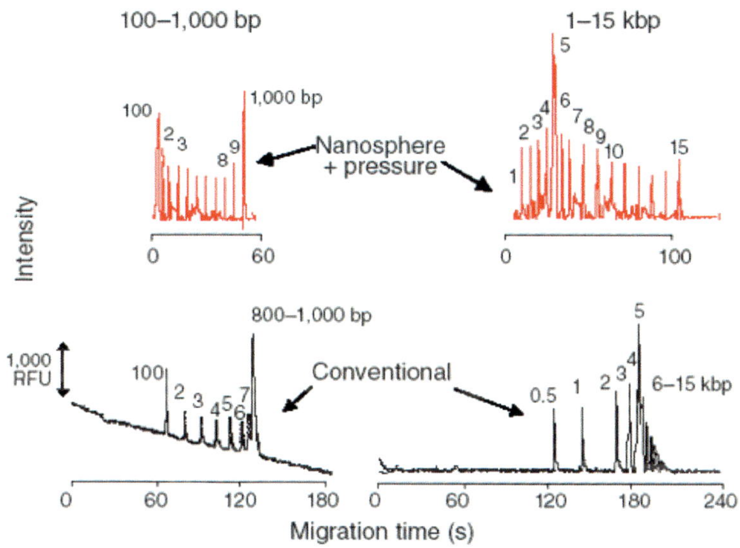

Figure 9.7 Effectiveness of DNA separations carried out using conventional polymers versus nanospheres with pressurization methods. Electropherograms are shown for the following samples. A 1 μg/ml solution of a 100 bp dsDNA ladder containing 10 fragments in a 1% (10 mg/ml) nanosphere solution (upper left). A 1 μg/mL solution of a 1 kbp dsDNA ladder containing 15 fragments nanosphere solution (upper right). A 100 bp dsDNA ladder in a conventional polymer (0.5% methylcellulose) solution using conventional electrophoresis (lower left). A 1 kbp dsDNA ladder in a conventional polymer (0.5% methylcellulose) solution using conventional electrophoresis (lower right). Adapted from M. Tabuchi, *et al.*, Nat. Biotech., 22, 337, 2004 and reproduced with permission.

To achieve such a high-speed and high-resolution separation, a complex procedure called pressurization is required subsequent to filling a microchannel with a 1% nanosphere solution. Instead of electrokinetic DNA sample loading and injection, an initial pressure about 2.5 kPa is applied to load the sample. This is followed by a secondary pressure application for 1 second. These pressurization processes are followed by DNA electrophoresis in the separation channel. Direct observation of DNA migration during the sample loading and injection process indicates the generation of a stacking effect of the packed nanospheres. Although methylcellulose solutions tend to lose their separation ability for large DNA molecules, nanosphere solutions can effectively sieve a wide range of DNA sizes and do not exhibit regions of length-independent electrophoretic mobility.

9.4.2 DNA Analysis by PEGylated-Latex Mixed Polymer Solution in a Chip

To investigate the effects of PEGylated-latex in a polymer solution, 80 nm PEGylated latex was mixed into a 0.7% hydroxypropyl methylcellulose (HPMC) solution and applied to microchip electrophoresis. Several DNA ladders were then separated.[12] As shown in Figure 9.8, a reduction in migration time was observed when using the PEGylated-latex mixed solution. Significant improvements in peak separation were also observed for a wide range of DNA sizes, ranging from 10 bp to 2 kbp. Nevertheless, since 80 nm latex without PEG modification is unable to separate DNA of all sizes, the high PEG density on the latex surface is likely to affect the separation performance.

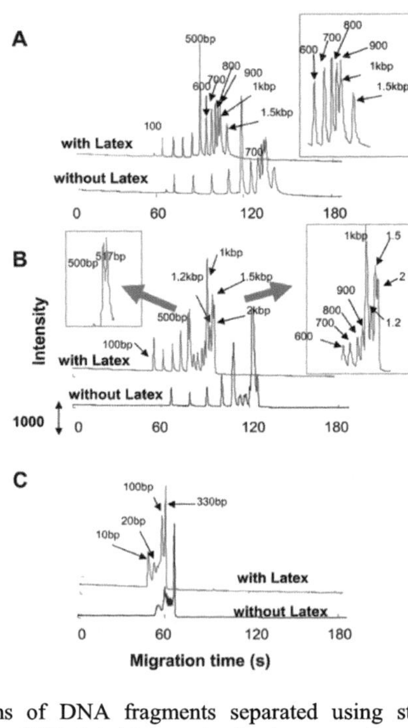

Figure 9.8 Electropherograms of DNA fragments separated using standard polymer medium (0.7% HPMC) and 80 nm-PEGylated-latex mixed polymer medium. (A) 100 bp DNA ladder (100 bp to 1500 bp), (B) 2 Long DNA ladder; the 500 bp band is a doublet of a 500 bp and 517 bp band, (C) 10 bp ladder. Each upper electropherogram has the 80 nm-PEGylated-latex mixed polymer medium and each lower electropherogram has 0.7% HPMC. Adapted from M. Tabuchi, *et al.*, Lab. Chip, 5, 199, 2005 and reproduced with permission.

9.4.3 Single DNA Molecule Imaging in a Nanoball solution

To further analyze the differences between DNA migration in nanosphere solutions and conventional polymer systems, the migration of single DNA molecules was visualized using fluorescence microscopy. DNA molecules exhibit intra-chain segregation, where folded compact regions and unfolded coil regions coexist.[26] This intra-chain segregated structure was observed in a 1% nanosphere solution under electrophoresis but never observed in conventional agarose gel or nanosphere-free solutions. As shown in Figure 9.9, it appears that the folded parts are connected by an unfolded coil structure. This may be due to the closely packed nanosphere solution which has a limited amount of extra space between nanospheres. Considering that DNA separations within dilute solutions of nanospheres (<1%) could not be achieved, this intra-chain segregated structure in the highly packed nanosphere solution is clearly essential in providing the separation ability. Although physical interaction between the folded part of DNA and the nanospheres might lead to separation under optimal nanosphere concentrations, a theoretical explanation for the separation mechanism has not yet been obtained.

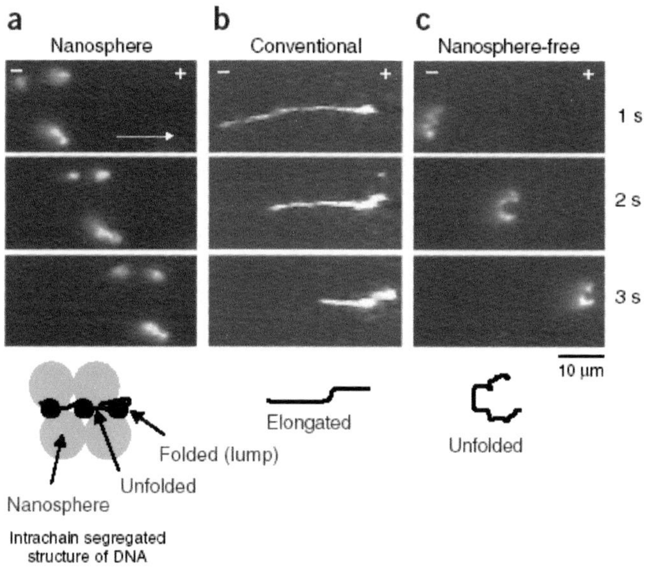

Figure 9.9 Visualization of single DNA molecules. (a,b) Sequential fluorescent images of DNA migration behaviour (T4 DNA, 165.6 kbp) under an electric field (10 V) in a 1% nanosphere solution (a), a conventional agarose gel (b) (1% agarose), and a control buffer containing no nanospheres (c). An illustration of typical DNA conformation is given at the bottom of each figure. Adapted from M. Tabuchi, *et al.*, Nat. Biotech., 22, 337, 2004 and reproduced with permission.

9.5 CONCLUSIONS

Two types of approach, "top-down" and "bottom-up", for DNA separation have been introduced and described. Although each approach possesses specific advantages and disadvantages for enhancing the analytical performance of chip-based electrophoresis devices, it seems that the combination of both approaches in an integrated device may define a potential route to further performance improvements. Furthermore highly controlled nanostructures should also provide precious information on DNA dynamics and allow the elucidation of some interesting theoretical predictions.

References

1. M.N. Albarghouthi, A.E. Barron, *Electrophoresis*, 2000, **21**, 4096-4111.
2. J. M. Meyer and G. S. Ginsburg, *Current Opinion in Chemical Biology*, 2002, **6**, 434-438.
3. W. D. Volkmuth and R. H. Austin, *Nature*, 1992, **358**, 600-602.
4. P.-G. de Gennes, *Scalling Concepts in Polymer Physics*, Cornell University Press, Ithaca, NY, 1979.
5. J. Han and H. G. Craighead, *Science*, 2000, **288**, 1026-1029.
6. J. Han and H. G. Craighead, *Anal Chem*, 2002, **74**, 394-401.
7. C. F. Chou, O. Bakajin, S. W. Turner, T. A. Duke, S. S. Chan, E. C. Cox, H. G. Craighead and R. H. Austin, *Proc Natl Acad Sci U S A*, 1999, **96**, 13762-13765.
8. M. Cabodi, Y. F. Chen, S. W. Turner, H. G. Craighead and R. H. Austin, *Electrophoresis*, 2002, **23**, 3496-3503.
9. P. S. Doyle, J. Bibette, A. Bancaud and J. L. Viovy, *Science*, 2002, **295**, 2237.
10. N. Kaji, Y. Tezuka, Y. Takamura, M. Ueda, T. Nishimoto, H. Nakanishi, Y. Horiike and Y. Baba, *Anal Chem*, 2004, **76**, 15-22.
11. M. Tabuchi, M. Ueda, N. Kaji, Y. Yamasaki, Y. Nagasaki, K. Yoshikawa, K. Kataoka and Y. Baba, *Nat Biotechnol*, 2004, **22**, 337-340.
12. M. Tabuchi, Y. Katsuyama, K. Nogami, H. Nagata, K. Wakuda, M. Fujimoto, Y. Nagasaki, K. Yoshikawa, K. Kataoka and Y. Baba, *Lab Chip*, 2005, **5**, 199-204.
13. N. Kaji, R. Ogawa, A. Oki, Y. Horiike, M. Tokeshi and Y. Baba, *Anal Bioanal Chem*, 2006, **386**, 759-764.
14. M. R. Mohamadi, L. Mahmoudian, N. Kaji, M. Tokeshi, H. Chuman and Y. Baba, in *Nanotoday*, 2006, pp. 38-45.
15. N. Kaji, A. Oki, R. Ogawa, Y. Takamura, T. Nishimoto, H. Nakanishi, Y. Horiike, M. Tokeshi and Y. Baba, *Israel Jounral of Chemistry*, 2007, **47**, 161-169.
16. L. Mahmoudian, M. R. Mohamadi, N. Kaji, M. Tokeshi and Y. Baba, in *Handbook of Capillary and Microchip Electrophoresis and Associated Microtechniques*, ed. J. P. Landers, CRC Press, 2007, pp. 1527-1542.
17. R. Ogawa, N. Kaji, S. Hashioka, Y. Baba and Y. Horiike, *Japanese Journal of Applied Physics*, 2007, **46**, 2771-2774.
18. T. Yasui, N. Kaji, R. Ogawa, S. Hashioka, M. Tokeshi, Y. Horiike and Y. Baba, in *The proceedings of micro Total Analysis Systems 2007*, Paris, 2007, pp. 1207-1209.
19. B.A. Buchholz, W. Shi, A.E. Barron, *Electrophoresis*, 2002, **23**, 1398-1409.
20. M. F. Huang, C. C. Huang and H. T. Chang, *Electrophoresis*, 2003, **24**, 2896-2902.
21. M. F. Huang, Y. C. Kuo, C. C. Huang and H. T. Chang, *Anal Chem*, 2004, **76**, 192-196.
22. K. Ito, H. Tsuchida, A. Hayashi, T. Kitano, E. Yamada and T. Matsumoto, *Polymer Journal*, 1985, **17**, 827-839.
23. R. Ogawa, Y. Nagasaki, N. Shibata, H. Otsuka and K. Kataoka, *Polymer Journal*, 2002, **34**, 868.

24. D. C. Schwartz and M. Koval, *Nature*, 1989, **338**, 520-522.
25. S. B. Smith, P. K. Aldridge and J. B. Callis, *Science*, 1989, **243**, 203-206.
26. K. Yoshikawa, *Adv Drug Deliv Rev*, 2001, **52**, 235-244.

Subject Index